Mohammed Shehata

Time-Optimal Control And Controllability Of n x n Parabolic Systems

Mohammed Shehata

Time-Optimal Control And Controllability Of n x n Parabolic Systems

LAP LAMBERT Academic Publishing

Impressum / Imprint

Bibliografische Information der Deutschen Nationalbibliothek: Die Deutsche Nationalbibliothek verzeichnet diese Publikation in der Deutschen Nationalbibliografie; detaillierte bibliografische Daten sind im Internet über http://dnb.d-nb.de abrufbar.

Alle in diesem Buch genannten Marken und Produktnamen unterliegen warenzeichen-, marken- oder patentrechtlichem Schutz bzw. sind Warenzeichen oder eingetragene Warenzeichen der jeweiligen Inhaber. Die Wiedergabe von Marken, Produktnamen, Gebrauchsnamen, Handelsnamen, Warenbezeichnungen u.s.w. in diesem Werk berechtigt auch ohne besondere Kennzeichnung nicht zu der Annahme, dass solche Namen im Sinne der Warenzeichen- und Markenschutzgesetzgebung als frei zu betrachten wären und daher von jedermann benutzt werden dürften.

Bibliographic information published by the Deutsche Nationalbibliothek: The Deutsche Nationalbibliothek lists this publication in the Deutsche Nationalbibliografie; detailed bibliographic data are available in the Internet at http://dnb.d-nb.de.

Any brand names and product names mentioned in this book are subject to trademark, brand or patent protection and are trademarks or registered trademarks of their respective holders. The use of brand names, product names, common names, trade names, product descriptions etc. even without a particular marking in this work is in no way to be construed to mean that such names may be regarded as unrestricted in respect of trademark and brand protection legislation and could thus be used by anyone.

Coverbild / Cover image: www.ingimage.com

Verlag / Publisher:
LAP LAMBERT Academic Publishing
ist ein Imprint der / is a trademark of
OmniScriptum GmbH & Co. KG
Heinrich-Böcking-Str. 6-8, 66121 Saarbrücken, Deutschland / Germany
Email: info@lap-publishing.com

Herstellung: siehe letzte Seite /
Printed at: see last page
ISBN: 978-3-659-34570-8

Time-Optimal Control

And

Controllability

Of

$n \times n$ Parabolic Systems

By

Mohammed Shehata

Contents

Chapter 1

Preliminaries

1.1 Introduction

[1]

Optimal control theory has been used in the solution of an enormous variety of problems in physics, engineering, economic and biology. Many of the problem of design in airframe, shipbuilding, electronic, and other engineering fields are in essence, problems of control.

To define a classical control problem, we require to describe the components of the problem. To start the definitions we need

(i) a real closed time interval $I = [t_0, T]$, with $t_0 < T$,

(ii) a bounded and closed subset U of $\Re^n$ that in which the control functions take values,

(iii) a differential equation describing the control system, satisfied by the trajectory function $t \in I \to y(t) \in \Re^n$, and a control function $t \in I \to u(t) \in U$,

(iv) an observation function $z(t, y(t), u(t))$ which is assumed to be known.

We can put further conditions on this function as necessary.

A classical optimal control problem is of finding an admissible control $u \in U$ which satisfies the differential equation describing the controlled system and minimizes the

---

[1]This thesis is written by LaTeX

4

cost functional $J(y, u) = \int_I f_0(t, u, y)dt$. The objective is to determine an admissible control, called *optimal control*, that provides a satisfactory state for us and that minimizes the value of functional J.

Perhaps the most widely studied type of problem in the mathematical theory of control is the "*time optimal*" control problem. The aim of control in this problem is to transform(or steering) any initial state of a dynamic system into a desired stationary state (or to hitting a target set)in minimal time. i.e., the cost function in this problem is the time in which a system is driving to desired state. Such problems are physical meaningful only if constraints are imposed on the control variables u, since otherwise, trivially, the desired state could be achieved in zero time by the application of controls of infinite amplitude.

The solution of the time-optimal control problem subject to bounded input is *bang-bang control*, i.e., the control in which the input variable takes either the maximum or minimum values. We mention the work of Wang[1], where a bang-bang principle of time optimal internal controls of heat equation was considered.

Time-optimal control has been studied extensively for "lumped parameter systems" i.e.,systems governed by ordinary differential equations. We refer the readers to classical books for Henry Herms and Joseph P.Lasalle [3] and Knowles[4]

In recent years, significant emphasis has been given to study the optimal control for systems described by partial differential equations, here referred to as distributed systems, arises whenever the spatial distribution of variables is to be controlled. Various optimization problems associated with the optimal control of distributed parameter systems have been studied recently by many authors, we mention only, Fattorini [5]-[14], Wang [15],Barbu [17], and others in, [18], [19], [20], [21]-[22],[27]-[30],[23],[24]-[25]and [26] besides the definite book "Optimal control of system governed by partial differential equations" written by J.L.Lions [32] which is itself based on his monumental joint work with Magenes [33]. In our M.Sc. thesis [35], and our paper [36]-[44], we discussed some time optimal control problems for some parabolic and hyperbolic systems.

In practical applications, the behavior of many dynamical systems which describes a state of time-optimal control problems depends upon their past histories. This phenomenon can be induced by the presence of time delays. Due to the inherent difficulties in solving control problems with time delays, the progress in this area has been slow. Here, we mention the work of Wang [16], where the time optimal control for a class of ordinary differential-difference equation with time lag was considered. Also, we mention the work of Knowles [2], where a Time optimal control of parabolic systems with boundary condition involving time delays was considered and it is shown that the optimal control is characterized in terms of an adjoint system and it is of the bang-bang type.

Controllability is a mathematical problem, which consists in determining the targets to which one can drive the state of some dynamical system, by means of a control parameter present in the equation. Many physical systems such as quantum systems, fluid mechanical systems, wave propagation, diffusion phenomena, etc. are represented by an infinite number of degrees of freedom, and their evolution follows some partial differential equation. Finding active controls in order to properly influence the dynamics of these systems generate highly involved problems. The control theory for PDEs, and among this theory, controllability problems, is a mathematical description of such situations. Any dynamical system represented by a PDE, and on which an external influence can be described, can be the object of a study from this point of view. In 1978, D.L. Russell [45] made a rather complete survey of the most relevant results that were available in the literature at that time. In that paper, the author described a number of different tools that were developed to address controllability problems, often inspired and related to other subjects concerning partial differential equations: multipliers, moment problems, nonharmonic Fourier series, etc. Various types of controllability of linear abstract dynamical systems defined in a Banach or Hilbert spaces have been recently extensively explored by several authors (see e.g.[46]-[62]). More recently, J.-L. Lions introduced the so called Hilbert Uniqueness Method (H.U.M.; see [31]).

Time-optimal control of distributed parameter systems governed by a system of parabolic equations is of special importance for the active control of structural systems for which the equations of motion are generally expressed by parabolic differential equations.In this work, we will focus our attention on some special aspects of minimum time problems for co-operative parabolic system involving Laplace operator with distributed or boundary control. In order to explain the results we have in mind, it is convenient to consider the abstract form:

Let V and H be two real Hilbert spaces such that V is a dense subspace of H. Identifying the dual of H with H, we may consider $V \subset H \subset V'$, where the embedding is dense in the following space. Let $A(t)$ ($t \in]0, T[$) be a family of continuous operators associated with a bilinear forms $\pi(t; ., .)$ defined on $V \times V$ which are satisfied Gårding's inequality

$$\pi(t; y, y) + c_0\|y\|_H^2 \geq c_1\|y\|_V^2, \quad c_0 \geq 0, \ c_1 > 0, \quad \text{for } y \in V, \ t \in [0, T]. \tag{1.1}$$

Then, from [32] and [33], for given f, y_0 and B be a bounded linear operator the following abstract systems:

$$\left. \begin{aligned} &\frac{d}{dt}y(t) + A(t)y(t) = f + Bu, \ t \in]0, T[, \\ &y(0) = y_0 \end{aligned} \right\} \tag{1.2}$$

have a unique solution, we denote it by $y(t; u)$. The time optimal control problem we shall concern reads:

$$min\{t : \ y(t; u) \in K, \ u \in U \}. \tag{1.3}$$

A control u^0 is called a time optimal control if $u^0 \in U$ and if there exists a number $\tau^0 > 0$ such that $y(\tau^0; u^0) \in K$ where $\tau^0 = \min\{\tau : \ y(\tau; u) \in K, \ u \in U \}$, we call the number τ^0 as the optimal time.

Three questions (problems) arise naturally in connection with this problem:

(a) Does there exist a control u, and $\tau > 0$ such that $y(\tau; u) \in K$? (this is an approximate controllability problem).

(b) Assume that the answer to (a) is in the affirmative. Does there exist a control u^0 which steering $y(\tau^0)$ to hitting a target set K in minimum time?

(c) If u^0 exists, is it unique? what additional properties does it have?

A typical application of a parabolic equation is the heat;

$$\frac{\partial y}{\partial t} = \Delta y + u \qquad \text{in } Q = \Omega \times]0, T[,$$

$$y(x, 0) = y_0(x) \quad \text{in } \Omega,$$

$$y(x, t) = 0 \qquad \text{on } \Gamma \times]0, T[,$$

$$(1.4)$$

where $\Omega \subset R^N$ is a bounded open domain with smooth boundary Γ, $\Delta = \sum_{k=1}^{N} \frac{\partial^2}{\partial x_k^2}$ is the Laplace operator and $y_0(x)$ is a given function in $L^2(\Omega)$.

The results in [63] partly overlap with results in [32] and they were shown that the system (1.4) (with U is closed convex subset of $L^2(Q)$ and $K = \{0\}$) is controllable and the corresponding time optimal control problem has at least one solution.

In the work [67], the authors gave a sufficient and necessary condition for the existence of time optimal control for the problem with the target set $K = \{0\}$ and certain controlled systems. These results will be stated as follows. Consider the following controlled system

$$\frac{\partial y}{\partial t} = \Delta y + ay + u \qquad \text{in } Q,$$

$$y(x, 0) = y_0(x) \quad \text{in } \Omega,$$

$$y(x, t) = 0 \qquad \text{on } \Sigma,$$

$$(1.5)$$

where a is a real number. Let $\{\lambda_i(a)\}_{i \geq 1}$, $\lambda_1(a) \leq \lambda_2(a) \dots$, be the eigenvalues of $-\Delta$ with the Dirichlet boundary condition and $\{e_i\}_{i \geq 1}$ be the corresponding eigenfunctions, which forms an orthogonal basis of $L^2(\Omega)$. We take the target set K to be the origin $\{0\}$ in $L^2(\Omega)$ and the control set U to be the set

$$U_\epsilon = \{u(., t) \in L^2(\Omega) : \|u\|_{L^2(\Omega)} \leq \epsilon \}$$

where ϵ is a positive number, namely, $U_\epsilon = B(0, \epsilon)$, the closed ball in $L^2(\Omega)$ centered at 0 and of radius ϵ. It was proved that the corresponding time optimal control problem has at least one solution if and only if $a \leq \lambda_1$.

More early, in the works [59], [68] and [69] the time optimal controls problem for globally controlled linear and semilinear parabolic equation with distributed or Neumann boundary controls were considered.

In this book, the time-optimal control problems(Besides the time-optimal control problem we study the controllability questions related to those

problems) for a regular parabolic system (such as heat equation) has been extended in the following different directions :

1. co-operative systems (coupled and $n \times n$ system)

2. several cases of observations

3. parabolic systems having delay

4. boundary controls

5. Infinite order parabolic system

1.2 Some Functional Spaces

In this section, we shall give some functional spaces which will be used later.

1.2.1 Interpolation Spaces

Let E_0 and E_1 be two Banach spaces continuously embedded into a Banach space E : $E_0 \subset E$, $E_1 \subset E$. such spaces are called an *interpolation* couple and is denoted by $\{E_0, E_1\}$ .Consider the Banach space

$$E_0 + E_1 := \{u = u_0 + u_1 : \ u_j \in E_j, \ j = 0, 1\}$$

$$\|u\|_{E_0+E_1} = \inf_{u=u_0+u_1, u_j \in E_j} (\|u_0\|_{E_0} + \|u_1\|_{E_1}).$$

Due to Triebel[64], the functional

$$K(t, u) := \inf_{u=u_0+u_1, u_j \in E_j} (\|u_0\|_{E_0} + t\|u_1\|_{E_1}), \quad u \in E_0 + E_1,$$

is continuous on $]0, T[$ in t , and the following estimate holds :

$$\min\{1, t\}\|u\|_{E_0+E_1} \le K(t, u) \le \max\{1, t\}\|u\|_{E_0+E_1}.$$

An *interpolation* space for $\{E_0, E_1\}$ by the K- method is defined as follows :

$$[E_0, E_1]_{\theta,p} = \{u \in E_0 + E_1 : \|u\|_{[E_0,E_1]_{\theta,p}} < \infty, \ 0 < \theta < 1, 0 \le p < \infty\},$$

$$\|u\|_{[E_0,E_1]_{\theta,p}} = \left[\int_0^T (t^{-1-\theta p} K^p(t,u)dt\right]^{\frac{1}{p}}$$

$$[E_0,E_1]_{\theta,\infty} = \{u \in E_0 + E_1 : \|u\|_{[E_0,E_1]_{\theta,\infty}} < \infty, \ 0 < \theta < 1,\}$$

$$\|u\|_{[E_0,E_1]_{\theta,\infty}} = \sup_{t\in]0,T[} t^{-\theta} K(t,u).$$

Let the embedding $E_0 \subset E_1$ be continuous . Consider the Banach space

$$W(0,T;E_0,E_1) := \left\{u : u \in L^2(0,T;E_0), u^{(1)} \in L^2(0,T;E_1)\right\}$$

with the norm

$$\|u\|_{W(0,T;E_0,E_1)} := \|u\|_{L^2(0,T;E_0)} + \|u^{(1)}\|_{L^2(0,T;E_1)}$$

1.2.2   Sobolev spaces

Let Ω be an open subset of $\Re^N$ with an infinitely differentiable boundary Γ . If $\alpha = (\alpha_1,\ldots,\alpha_N)$ is a N-dimensional vector of integers $\alpha_j \geq 0$,then

$$D^\alpha = D_1^{\alpha_1} D_2^{\alpha_2} \ldots D_N^{\alpha_N}; \quad \text{the order of the derivative is }, \ \mid \alpha \mid = \sum_{i=1}^N \alpha_i$$

and

$$D_k = \frac{\partial}{\partial x_k}, \quad 1 \leq k \leq N.$$

Then, the $W_q^m(\Omega)$ is a Banach space of functions $u(x)$ that have generalized derivatives on Ω up to $m-$ order inclusive , for which the following norm is finite:

$$\|u\|_{W_q^m(\Omega)} = \left[\sum_{|\alpha|\leq m} \|D^\alpha u\|_{L^q(\Omega)}^q\right]^{\frac{1}{q}}.$$

Let s_0 and s_1 be non -negative integers, $0 < \theta < 1$, $1 \leq q \leq \infty$, $1 \leq p \leq \infty$ and $s = (1-\theta)s_0 + \theta s_1$ .From Triebl[64] it follows that if $s = (1-\theta)s_0 + \theta s_1 = (1-\theta')s_0' + \theta's_1'$, then ,

$$[W_p^{s_0}(\Omega), W_p^{s_1}(\Omega)]_{\theta,q} = [W_p^{s_0'}(\Omega), W_p^{s_1'}(\Omega)]_{\theta',q}$$

Consider the space

$$B_{p,q}^s(\Omega) := [W_p^{s_0}(\Omega), W_p^{s_1}(\Omega)]_{\theta,q},$$

where s_0, s_1 are non -negative integers, $0 < \theta < 1$, $1 \leq q \leq \infty$, $1 \leq p \leq \infty$ and $s = (1 - \theta)s_0 + \theta s_1$ .For s positive and not an integer , set

$$W_p^s(\Omega) = B_{p,p}^s(\Omega) := [W_p^{s_0}(\Omega), W_p^{s_1}(\Omega)]_{\theta,p}.$$

We denote by $\mathbf{D}(\Omega)$ the space of all test functions on Ω (infinitely differentiable functions with compact support contained in Ω) and by $W_{q,0}^m(\Omega)$,the closure of $\mathbf{D}(\Omega)$ in $W_q^m(\Omega)$ .Since the space $\mathbf{D}(\Omega)$ is dense in $L^p(\Omega)$ Adams [70], so is $W_{q,0}^m(\Omega)$, a fortiori $W_q^m(\Omega)$.

For $p = 2$ the alternate notation

$$W_2^m(\Omega) = H^m(\Omega), \ W_{2,0}^m(\Omega) = H_0^m(\Omega)$$

are Hilbert spaces equipped with the inner product

$$\big(y, z\big)_{H^m(\Omega)} = \sum_{|\alpha| \leq m} (D^\alpha y, D^\alpha z)_{L^2(\Omega)} .$$

and for s positive and not an integer , set

$$W_2^s(\Omega) = H^s(\Omega) := [H^{s_0}(\Omega), H^{s_1}(\Omega)]_{\theta,2}.$$

where s_0, s_1 are non -negative integers, $0 < \theta < 1$, and $s = (1 - \theta)s_0 + \theta s_1$

Let $1 \leq p < \infty$ .Let p' be the conjugate exponent of p , i.e $\frac{1}{p} + \frac{1}{p'} = 1$. The *dual* of the space $W_{p,0}^m(\Omega)$ where $m \geq 1$ is an integer is denoted by $W_{p'}^{-m}(\Omega)$. If $p = 2$, $H^{-m}(\Omega)$ is the dual of the space $H_0^m(\Omega)$ .and we have the following chain :

$$H_0^m(\Omega) \subset L^2(\Omega) \subset H^{-m}(\Omega).$$

$W_q^m(0, T; E)$, $1 \leq q < \infty$, with m integer , denotes a Banach space of functions $u(x)$ with values from E which have generalized derivatives up to $m-$ th order, inclusive , on $]0, T[$ and the following norm is finite.

$$\|u\|_{W_q^m(0,T;E)} := \sum_{k=0}^m \left[\int_0^T \|u^{(k)}(x)\|_E^q dx \right]^{\frac{1}{q}}$$

For $m = 0$ the alternate notation

$$W_q^0(0, T; E) = L^q(0, T; E)$$

and for $q = 2$ the alternate notation

$$W_2^m(0, T; E) = H^m(0, T; E)$$

is Hilbert space equipped with the inner product

$$\left(u, v\right)_{H^m(0,T;E)} = \sum_{k=0}^{m} \left(u^{(k)}, v^{(k)}\right)_{L^2(0,T;E)}.$$

.For s positive and not an integer , set

$$W_p^s(0, T; E) := [W_p^{s_0}(0, T; E), W_p^{s_1}(0, T; E)]_{\theta, p}.$$

and

$$H^s(0, T; E) := [H^{s_0}(0, T; E), H^{s_1}(0, T; E)]_\theta.$$

where s_0 , s_1 are non -negative integers, $0 < \theta < 1$, $1 \leq p \leq \infty$ and $s = (1 - \theta)s_0 + \theta s_1$

Let $]0, T[$ be a given finite time interval .Let $H^r(\Omega)$, $r \geq 0$, denote the Sobolev space of order r on Ω . For any pair of real numbers , $r, s \geq 0$, th e Sobolev space (Lions and Magenes[33])$H^{r,s}(Q)^n$ is defined by

$$H^{r,s}(Q)^n = H^0(0, T; (H^r(\Omega))^n) \cap H^s(0, T; (H^0(\Omega))^n), \quad Q = \Omega \times]0, T[$$

which is a Hilbert space with the norm :

$$\left(\int_0^T \|y(t)\|_{(H^r(\Omega))^n}^2 + \|y\|_{H^s(0,T;(H^0(\Omega))^n)}^2\right)^{\frac{1}{2}}$$

where $H^s(0, T; (H^0(\Omega))^n)$ denotes the Sobolev space of order s of vector functions defined on $]0, T[$ and taking values in $(H^0(\Omega))^n = (L^2(\Omega))^n$.

For example :

$$H^{2,1}(Q)^n = H^0(0, T; (H^2(\Omega))^n) \cap H^1(0, T; (H^0(\Omega))^n)$$

$$= L^2(0, T; (H^2(\Omega))^n) \cap H^1(0, T; (L^2(\Omega))^n)$$

$$= \left\{ \mathbf{y} = (y_k)_{k=1}^n : y_k, \frac{\partial y_k}{\partial x_i}, \frac{\partial^2 y_k}{\partial x_i \partial x_j}, \frac{\partial y_k}{\partial t} \in L^2(Q) \right\}.$$

and it's norm given by

$$\left(\int_Q \left[|\mathbf{y}|^2 + \sum_{i=1}^N \left(\frac{\partial \mathbf{y}}{\partial x_i} \right)^2 + \left(\frac{\partial^2 \mathbf{y}}{\partial x_i \partial x_j} \right)^2 + \left(\frac{\partial \mathbf{y}}{\partial t} \right)^2 \right] dx dt \right)^{\frac{1}{2}}$$

$$= \left(\sum_{k=1}^n \int_Q \left[|y_k|^2 + \sum_{i=1}^N \left(\frac{\partial y_k}{\partial x_i} \right)^2 + \left(\frac{\partial^2 y_k}{\partial x_i \partial x_j} \right)^2 + \left(\frac{\partial y_k}{\partial t} \right)^2 \right] dx dt \right)^{\frac{1}{2}}$$

1.2.3 Sobolev spaces of infinite order

The object of this subsection is to give the definition of *infinite order* Sobolev space $W_0^\infty\{a_\alpha, p\}$ where $a_\alpha \geq 0$ is a numerical sequence , $p \geq 1$ and the chain of the constructed spaces which will be used later .

For a bounded open subset Ω of $\Re^N$ with smooth boundary Γ ,We define the Sobolev space of infinite order as follows: ,

$$W_0^\infty\{a_\alpha, p\} = \left\{ \phi(x) \in C_0^\infty(\Omega) : \|\phi\|_{W_0^\infty\{a_\alpha, p\}} = \sum_{|\alpha|=0}^\infty a_\alpha \| D^\alpha \phi \|_{L^p(\Omega)}^p < \infty \right\}$$

where

$$C_0^\infty(\Omega) = \{\phi(x) \in C^\infty(\Omega) : \ \mathrm{supp}\phi \in \bar{\Omega}, \ \ \bar{\Omega} = \Omega \times \Gamma\}$$

The space $W_0^\infty\{a_\alpha, p\}$ is non -trivial if there exists a function $\phi \neq 0$, $\phi \in C_0^\infty(\Omega)$ such that

$$\|\phi\|_{W_0^\infty\{a_\alpha, p\}} = \sum_{|\alpha|=0}^\infty a_\alpha \| D^\alpha \phi \|_{L^p(\Omega)}^p < \infty$$

The question of non-triviality of these space is studied, for instance by Dubiniskii (see [75]-[78]).

The space $W_0^{-\infty}\{a_\alpha, p'\}$,where $\frac{1}{p} + \frac{1}{p'} = 1$,is defined as the formal conjugate space to the space $W_0^\infty\{a_\alpha, p\}$, namely

$$W_0^{-\infty}\{a_\alpha, p'\} := \left\{ \psi(x) : \psi(x) = \sum_{|\alpha|=0}^\infty a_\alpha D^\alpha \psi_\alpha(x) \right\}$$

where

$$\psi_\alpha \in L^{p'}(\Omega) \quad \text{and} \quad \sum_{|\alpha|=0}^\infty a_\alpha \| \psi_\alpha \|_{L^{p'}(\Omega)}^{p'} < \infty$$

The duality pairing of the spaces $W_0^\infty\{a_\alpha, p\}$ and $W_0^{-\infty}\{a_\alpha, p'\}$ is postulated by the formula

$$(\phi, \psi) = \sum_{|\alpha|=0}^{\infty} a_\alpha \int_\Omega \psi_\alpha D^\alpha \phi(x) dx,$$

where $\phi \in W_0^\infty\{a_\alpha, p\}$ and $\psi \in W_0^{-\infty}\{a_\alpha, p'\}$.

Now ,we consider the space $W_0^\infty\{a_\alpha, 2\}(\Omega)$ which is Hilbert space equipped with the inner product

$$\left(y, z\right)_{W_0^\infty\{a_\alpha, 2\}(\Omega)} = \sum_{|\alpha|=0} a_\alpha \left(D^\alpha y, D^\alpha z\right)_{L^2(\Omega)}.$$

An imbedding criterion is established .In this case $W_0^\infty\{a_\alpha, 2\}(\Omega)$ is every where dense in $L^2(\Omega)$ with topological inclusions , and $W_0^{-\infty}\{a_\alpha, 2\}(\Omega)$ denotes the topological dual space with respect to $L^2(\Omega)$ and we have the following inclusions ,

$$W_0^\infty\{a_\alpha, 2\} \subseteq L^2(\Omega) \subseteq W_0^{-\infty}\{a_\alpha, 2\}$$

We take the n-Cartesian product to the above Hilbert spaces.Then as above subsection we have the following chains

$$(W_0^\infty\{a_\alpha, 2\})^n \subseteq (L^2(\Omega))^n \subseteq (W_0^{-\infty}\{a_\alpha, 2\})^n.$$

1.3 Evolution Problem of First Order

In this section , we introduce the existence and uniqueness of the solution of evolution problem of first order (see Lions[32]ChapterIII).

1.3.1 Problem formulation

Let V and H be two Hilbert spaces with $V \subset H$, V dense in H with continuous injection . We identify H with its dual H'; we have

$$V \subset H \subset V',$$

Let t be the time variable , we assume that $t \in]0, T[$, $T < \infty$. A family of a bi-linear forms on V is given ;

$$u, v \rightarrow \pi(t; u, v) \quad \text{for each} \quad t \in]0, T[.$$

Relative to this family we assume,

$$\left. \begin{array}{l} \forall u, v \in V \text{ the functions} \quad t \rightarrow \pi(t; u, v) \text{ is measurable on} \\ ]0, T[\quad \text{and} \quad \left| \pi(t; u, v) \right| \leq c \|u\|_V \|v\|_V, \end{array} \right\} \tag{1.6}$$

and

$$\left. \begin{array}{l} \text{there exist} \quad \lambda \quad \text{such that} \quad \forall u \in V, \quad t \in]0, T[\\ \pi(t; u, u) + \lambda \|u\|_H \geq \alpha \|u\|_V^2, \quad \alpha > 0 \end{array} \right\} \tag{1.7}$$

For each t we may write

$$\pi(t; u, v) = < A(t)u, v >, \quad A(t)u \in V' \tag{1.8}$$

the bracket denoting the scalar product between V and V'.

We now define the space $L^2(0, T; V)$ as the space of (classes of) measurable functions

$$t \rightarrow f(t) \text{ of }]0, T[\rightarrow V$$

such that

$$\left(\int_0^\infty \|f(t)\|_V^2 dt \right)^{\frac{1}{2}} = \|f\|_{L^2(0,T;V)} < \infty$$

in the same manner we define $L^2(0, T; V')$ and we note ,

$$A(t) \in L\big(L^2(0, T; V); L^2(0, T; V') \big) \tag{1.9}$$

that is , if $f \in L^2(0, T; V')$, $A(t)f$ is the function $t \rightarrow A(t)f(t) \in V'$

It may be verified that this function is measurable and satisfies

$$\|A(t)f(t)\|_{V'} \leq c \|f(t)\|_V \quad (\text{from } (1.6))$$

whence (1.9)

We introduce the Hilbert space $W(0, T; V, V')$ (as in subsection 1.1 with $E_0 = V$, $E_1 = V'$)

We now Consider an evolution problem: Find $y \in W(0, T; V, V')$ such that

$$\left.\begin{array}{c} \dfrac{d}{dt}y(t) + A(t)y(t) = f, \quad f \text{ given in } L^2(0, T, V'), \\[2ex] \text{and} \quad y(0) = y_0, \quad y_0 \text{ given in H} \end{array}\right\} \tag{1.10}$$

We shall prove :

Theorem 1.3.1. *Under the hypotheses* (1.6),(1.7),*problem* (1.10) *admits a unique solution in* $W(0, T; V, V')$. *Furthermore the solution depends continuously on the data ; the bilinear map ,*

$$\{f, y_0\} \to y$$

is continuous from

$$L^2(0, T; V') \times H \to W(0, T; V, V')$$

Let us remark that:

Remark 1.3.1. *Lions -Magenes [33]. All functions* $f \in W(0, T; V, V')$ *are ,with eventual modifications on a set of zero, continuous from* $[0, T] \to H$.

Remark 1.3.2.

$$\left.\begin{array}{l} if T < \infty, \quad we \ may \ always \ reduce \ the \ problem \ to \ a \ case \ where \\[1ex] (1.7) \quad holds \ with \ \lambda = 0. \end{array}\right\} \tag{1.11}$$

In fact, if we set

$$y = \exp(kt)z,$$

problem (1.10) *is equivalent to*

$$(A(t) + kI)z + \frac{dz}{dt} = \exp(-kt)f, \quad z(0) = y_0,$$

and hence we have replaced $A(t)$ *by* $A(t) + kI$.*Choosing* $k = \lambda$,(1.11) *is obtained.*

We shall now prove Theorem1.3.1 assuming (1.11) holds

1.3.2 Proof of uniqueness

Let y satisfy (1.10) with $f = 0$, $y_0 = 0$. In the first equation in (1.10),let us take the scalar product with $y(t)$ under the duality between V and V'. We obtain

$$\pi(t; y(t), y(t)) + \left\langle \frac{dy(t)}{dt}, y(t) \right\rangle = 0.$$

But we may verify (from remark1.3.1) that ,

$$\int_0^T \left\langle \frac{dy(t)}{dt}, y(t) \right\rangle dt = \frac{1}{2}\|y(T)\|_H^2 \quad \text{(since } y(0) = 0\text{)},$$

and hence

$$\int_0^T \pi(t; y(t), y(t))dt + \frac{1}{2}\|y(T)\|_H^2 = 0.$$

Therefore using (1.11)

$$\alpha \int_0^T \|y(t)\|_V^2 dt + \frac{1}{2}\|y(T)\|_H^2 \leq 0,$$

and hence $y = 0$.

1.3.3   Proof of existence

To simplify the exposition somewhat, let us assume that V is separable .Therefore there exists a countable set which is dense in V . We may then find (in an infinity of ways) a " basis" $w_1, w_2, \ldots, w_m, \ldots$ in V in the following sense:

$\forall m,\ w_1, \ldots, w_m$ are linearly independent and the linear combinations

$$\sum_{\text{finite}} \xi_j w_j, \quad \xi_j \in \Re, \text{ are dense in V.}$$

Let us define an " approximate solution " of (1.10) by

$$y_m(t) = \sum_{i=1}^m g_{im} w_i,$$

where the g_{im} are chosen such that

$$\pi(t; y_m(t), w_j) + \left\langle \frac{dy_m(t)}{dt}, w_j \right\rangle = \langle f(t), w_j \rangle, \quad 1 \leq j \leq m \qquad (1.12)$$

and

$$y_m(0) = y_{0m} = \sum_{i=1}^m \xi_{im} w_i, \quad \sum_{i=1}^m \xi_{im} w_i \to y_0 \text{ in } H \text{ as } m \to \infty \qquad (1.13)$$

System (1.12),(1.13) is a system of m linear differential equations in g_{im} of the form

$$W_m \frac{d}{dt} g_m + A_m(t) g_m = f_m \quad , g_m(0) = (\xi_{im})$$

where

$$W_m = | <w_i, w_j> |, \qquad\qquad A_m = |\pi(t; w_i, w_j)|.$$
$$g_m(t) = (g_{im})_{i=1}^m \qquad\qquad f_m(t) = (<f(t), w_j>).$$

Since $\det W_m \neq 0$ problem (1.12),(1.13) admits a unique solution. We shall show that as $m \to \infty$, $y_m \to y$, y being a solution of (1.10) .Multiplying (1.12) by $g_{jm}(t)$ and summing over j , we obtain

$$\pi(t; y_m(t), y_m(t)) + \left\langle \frac{dy_m(t)}{dt}, y_m(t) \right\rangle = <f(t), y_m(t)>,$$

that is ,

$$\pi(t; y_m(t), y_m(t)) + \frac{1}{2}\frac{d}{dt}\|y_m(t)\|_H^2 = <f(t), y_m(t)>,$$

whence using(1.10),

$$\|y_m(T)\|_H^2 + 2\alpha \int_0^T \|y_m(t)\|_V^2 dt \leq \|y_{0m}\|_H^2 + 2\int_0^T |<f(t), y_m(t)>| dt$$

$$\leq \|y_{0m}\|_H^2 + 2\int_0^T \|f(t)\|_{V'}\|y_m(t)\|_V dt$$

$$\leq \|y_{0m}\|_H^2 + \int_0^T \|y_m(t)\|_V^2 dt$$

$$+ \frac{1}{\alpha}\int_0^T \|f(t)\|_{V'}^2 dt.$$

Finally (since $\|y_{0m}\|_H \geq c\|y_0\|_H$, ,cf(1.13)),

$$\int_0^T \|y_m(t)\|_V^2 dt \leq c\left(\|y_0\|_H^2 + \int_0^T \|f(t)\|_{V'}^2 dt \right) \qquad (1.14)$$

Therefore y_m ranges in a bounded set in $L^2(0,T;V)$ and we may extract a subsequence y_μ such that

$$y_\mu \to z \quad \text{weakly in } L^2(0,T;V). \qquad (1.15)$$

Let j be fixed but arbitrary and let $\mu > j$ .Then (1.12) is valid with $m = \mu$. Multiply both sides of (1.12)by $\phi(t)$ where

$$\phi(t) \in C^1[0,T], \quad \phi(T) = 0 \qquad (1.16)$$

and integrate over $]0,T[$ .Setting $\phi_j(t) = \phi(t)w_j$, we have,

$$\int_0^T \left[\pi(t; y_\mu(t), \phi_j(t)) - \left\langle y_\mu(t), \frac{d\phi_j}{dt} \right\rangle \right] dt =$$
$$= \int_0^T <f(t), \phi_j(t)> dt + <y_{0\mu}, \phi_j(0)>. \qquad (1.17)$$

By virtue of (1.15) ,we may pass to the limit in (1.17).We then have,

$$\int_0^T \left[\pi(t; z, \phi_j) - \left\langle z, \frac{d\phi_j}{dt} \right\rangle \right] dt = \int_0^T <f, \phi_j> dt+ <y_0, \phi_j(0) > . \qquad (1.18)$$

But the above is true for any ϕ satisfying (1.16).Therefore we may take $\phi \in \mathbf{D}(]0, T[)$ and hence (1.18)gives

$$\frac{d}{dt} < z(t), w_j > +\pi(t; z(t), w_j) =< f(t), w_j > \qquad (1.19)$$

where the derivative is taken in $\mathbf{D}'(]0, T[)$. But in $(1.19) j$ is arbitrary and since finite linear combination of w_j are dense in V ,we deduce

$$\frac{dz}{dt} + A(t)z = f \qquad (1.20)$$

Therefore ,

$$\frac{dz}{dt} = f - A(t)z \in L^2(0, T; V') \quad \text{and hence } z \in W(0, T; V, V').$$

This enables us to integrate by parts in t .Then taking into account (1.20)we option

$$< z(0), w_j > \phi(0) =< y_0, w_j > \phi(0) \quad \forall j, \ \forall \phi.$$

Hence

$$< z(0), w_j >=< y_0, w_j > \quad \text{and thus } z(0) = y_0.$$

Hence z is a solution and using $(1.10) z = y$ is a solution . We may then replace (1.15)by

$$y_\mu \to y \quad \text{weakly in } L^2(0, T; V).$$

The estimate (1.14) gives us

$$\int_0^T \|y(t)\|_V^2 dt \le c \left(\|y_0\|_H^2 + \int_0^T \|f(t)\|_{V'}^2 dt \right) \qquad (1.21)$$

Further ,since $\frac{dy}{dt} = f - A(t)y$ we have with (1.21)

$$\|\frac{dy}{dt}\|_{L^2(0, T; V')}^2 \le c' \left(\|y_0\|_H^2 + \int_0^T \|f(t)\|_{V'}^2 dt \right).$$

1.4 Some Parabolic Systems

1.4.1 Parabolic system: involving Laplace operator

In this subsection we consider the following parabolic system involving Laplace operator .

$$
\left.
\begin{aligned}
\frac{\partial y}{\partial t} - \Delta y = f, \quad f \in L^2(Q) \qquad &\text{,in } Q = \Omega \times]0, T[, \\
y(x, 0) = y_0(x), \quad y_0(x) \in L^2(\Omega), \qquad &\text{in } \Omega, \\
y_(x, t) = 0 \qquad &\text{,on } \Sigma = \Gamma \times]0, T[.
\end{aligned}
\right\} \tag{1.22}
$$

where $\Omega \subset R^N$ is a bounded open domain with smooth boundary Γ .

We shall use a chain of the form

$$
H_0^1(\Omega) \subseteq L^2(\Omega) \subseteq H_0^{-1}(\Omega)
$$

For $y, \phi \in H_0^1(\Omega)$ and $t \in]0, T[$ let us define a family of continues bilinear forms

$$
\pi(t; ., .) : H_0^1(\Omega) \times H_0^1(\Omega) \to \Re
$$

by :

$$
\begin{aligned}
\pi(t; y, \phi) = \pi(y, \phi) &= \sum_{i=1}^{N} \int_\Omega \frac{\partial y}{\partial x_i} \frac{\partial \phi}{\partial x_i} dx = \sum_{i=1}^{N} \int_\Omega \frac{\partial^2 y}{\partial x_i^2} dx \\
&= \int_\Omega (\Delta y) \phi dx = \langle \Delta y, \phi \rangle_{L^2(\Omega)}
\end{aligned}
$$

The above continous bilinear form satisfies the coercive condition

$$
\left.
\begin{aligned}
\text{there exist} \quad \lambda \quad \text{such that} \quad \forall y \in H_0^1(\Omega), \quad t \in]0, T[\\
\pi(t; y, y) + \lambda \|y\|_{L^2(\Omega)} \geq \alpha \|y\|_{H_0^1(\Omega)}, \quad \alpha > 0 (\text{ take } \lambda = 2, \ \alpha = 1)
\end{aligned}
\right\}
$$

Then from Theorem 1.1, there exists a unique solution

$$
y \in W(0, T; (H_0^1(\Omega)); H_0^{-1}(\Omega))
$$

satisfying the Dirichlet problem (1.22)

1.4.2 Parabolic system: involving Operator of infinite order

In this subsection, we consider a parabolic system involving Operator of infinite order.

We shall use a chain of the form

$$W_0^\infty\{a_\alpha, 2\} \subseteq L^2(\Omega) \subseteq W_0^{-\infty}\{a_\alpha, 2\}$$

where $\Omega \subset R^N$ is a bounded open domain with smooth boundary Γ

For $y, \phi \in W_0^\infty\{a_\alpha, 2\}$ and $t \in]0, T[$ let us define a family of continues bilinear forms

$$\pi(t; ., .) : W_0^\infty\{a_\alpha, 2\} \times W_0^\infty\{a_\alpha, 2\} \to \Re$$

by :

$$\pi(t; \mathbf{y}, \varphi) = \pi(\mathbf{y}, \varphi) = \int_\Omega \sum_{|\alpha|=0}^\infty a_\alpha (D^\alpha y(x))(D^\alpha \varphi(x)) dx$$

$$= \int_\Omega \left(\sum_{|\alpha|=0}^\infty a_\alpha (-1)^\alpha D^{2\alpha} y(x) \right) \varphi(x) dx$$

$$= \left\langle \left(\sum_{|\alpha|=0}^\infty a_\alpha (-1)^\alpha D^{2\alpha} y(x) \right), \varphi(x) \right\rangle_{L^2(\Omega)}$$

where $y, \varphi \in W_0^\infty\{a_\alpha, 2\}$ and $D_k y(x) = \frac{\partial y(x)}{\partial x_k}$

The above continues bilinear form satisfies the coercive condition

$$\left. \begin{array}{c} \text{there exist} \quad \lambda \quad \text{such that} \quad \forall y \in W_0^\infty\{a_\alpha, 2\}, \quad t \in]0, T[\\ \pi(t; y, y) + \lambda \|y\|_{L_2(\Omega)} \geq \alpha \|y\|_{W_0^\infty\{a_\alpha, 2\}}, \quad \alpha > 0(\text{ take } \lambda = 2, \ \alpha = 1) \end{array} \right\}$$

Then from Theorem 1.1,there exists a unique solution

$$y \in W(0, T; W_0^\infty\{a_\alpha, 2\}; W_0^{-\infty}\{a_\alpha, 2\}$$

satisfying the following Dirichlet problem .

$$\left. \begin{array}{ll} \dfrac{\partial y}{\partial t} + \left(\displaystyle\sum_{|\alpha|=0}^\infty a_\alpha (-1)^\alpha D^{2\alpha} \right) y = f, \quad f \in L^2(Q) & \text{,in } Q = \Omega \times]0, T[, \\[4mm] y(x, 0) = y_0(x), \quad y_0(x) \in L^2(\Omega) & \text{,in } \Omega, \\[2mm] y(x, t) = 0 & \text{,on } \Sigma = \Gamma \times]0, T[. \end{array} \right\}$$

1.4.3 Coupled systems

In this subsection we consider the following coupled system involving Laplace operators.

$$
\left.
\begin{aligned}
\frac{\partial y_1}{\partial t} - \Delta y_1 + y_1 - y_2 &= f_1, \quad f_1 \in L^2(Q) && \text{,in } Q, \\
\frac{\partial y_2}{\partial t} + y_1 - \Delta y_2 + y_2 &= f_2, \quad f_2 \in L^2(Q) && \text{,in } Q, \\
y_1(x,0) = y_{1,0}(x),, \quad y_{1,0}(x) &\in L^2(\Omega) && \text{,in } \Omega, \\
y_2(x,0) = y_{2,0}(x), \quad y_{2,0}(x) &\in L^2(\Omega) && \text{,in } \Omega, \\
y_1 = y_2 &= 0 && \text{,on } \Sigma.
\end{aligned}
\right\} \quad (1.23)
$$

We shall use a chain of the form

$$
(H_0^1(\Omega))^2 \subseteq (L^2(\Omega))^2 \subseteq (H^{-1}(\Omega))^2
$$

For $\mathbf{y} = (y_1, y_2)$, $\phi = (\phi_1, \phi_2) \in (H_0^1(\Omega))^2$ and $t \in]0, T[$ let us define a family of continues bilinear forms

$$
\pi(t; ., .) : (H_0^1(\Omega))^2 \times (H_0^1(\Omega))^2 \to \Re
$$

by :

$$
\begin{aligned}
\pi(t; y, \phi) &= \pi(y, \phi) \\
&= \sum_{i=1}^{N} \int_{\Omega} \left[\frac{\partial y_1}{\partial x_i} \frac{\partial \phi_1}{\partial x_i} + \frac{\partial y_2}{\partial x_i} \frac{\partial \phi_2}{\partial x_i} \right] dx + \int_{\Omega} [y_1 \phi_1 + y_1 \phi_2 - y_2 \phi_1 + y_2 \phi_2] \, dx \\
&= \sum_{i=1}^{N} \int_{\Omega} \left[\frac{\partial^2 y_1}{\partial x_i} \phi_1 + \frac{\partial^2 y_2}{\partial x_i} \phi_2 \right] dx + \int_{\Omega} [y_1 \phi_1 + y_1 \phi_2 - y_2 \phi_1 + y_2 \phi_2] \, dx \\
&= \int_{\Omega} [(\Delta y_1) \phi_1 + (\Delta y_2) \phi_2] \, dx + \int_{\Omega} [y_1 \phi_1 + y_1 \phi_2 - y_2 \phi_1 + y_2 \phi_2] \, dx \\
&= \int_{\Omega} [(\Delta y_1) + y_1 - y_2] \phi_1 + \int_{\Omega} [(\Delta y_2) + y_1 + y_2] \phi_2 dx \\
&= \langle [(\Delta y_1) + y_1 - y_2], \phi_1 \rangle_{L^2(\Omega)} + \langle [(\Delta y_2) + y_1 + y_2], \phi_2 \rangle_{L^2(\Omega)} \\
&= \langle \mathbf{A}\mathbf{y}, \phi \rangle_{(L^2(\Omega))^2}
\end{aligned}
$$

where

$$
\mathbf{A} \begin{pmatrix} y_1 \\ y_2 \end{pmatrix} = \begin{pmatrix} -\Delta & -I \\ I & -\Delta \end{pmatrix} \begin{pmatrix} y_1 \\ y_2 \end{pmatrix}
$$

The above continues bilinear form is coercive on $(H_0^1(\Omega))^2$:

$$\pi(\mathbf{y}, \mathbf{y}) = \|\mathbf{y}\|^2 = \langle \mathbf{Ay}, \mathbf{y} \rangle_{(L^2(\Omega))^2}$$

Then from Theorem 1.1,there exists a unique solution

$$y \in W(0, T; (H_0^1(\Omega))^2; (H_0^{-1}(\Omega))^2)$$

satisfying the Dirichlet problem (1.23)

Remark 1.4.1. *the parabolic system* (1.23) *can be written in the following operator model*

$$\left.\begin{array}{ll} \dfrac{\partial}{\partial t} \begin{pmatrix} y_1 \\ y_2 \end{pmatrix} + \mathbf{A} \begin{pmatrix} y_1 \\ y_2 \end{pmatrix} = \begin{pmatrix} f_1 \\ f_2 \end{pmatrix}, & in \ \ Q, \\[3mm] \begin{pmatrix} y_1(x, 0) \\ y_2(x, 0) \end{pmatrix} = \begin{pmatrix} y_{1,0}(x) \\ y_{2,0}(x) \end{pmatrix} & in \ \ \Omega, \\[3mm] \begin{pmatrix} y_1 \\ y_2 \end{pmatrix} = \begin{pmatrix} 0 \\ 0 \end{pmatrix} & on \ \ \Sigma. \end{array}\right\}$$

Chapter 2

Time Optimal Control of First Order Evolution Systems

This chapter is devoted to presented some results concerning with time optimal control of first order evolution systems. We formulate the time optimal control problem in the case of a target set, then we study existence of the time optimal control and some of its properties. Finally we give some mathematical examples for parabolic systems (2×2 parabolic systems) involving different type of operators (Laplace operator - operator of infinite order).

2.1 Problem statement

We consider the system whose state is given by

$$\left. \begin{aligned} \frac{dy}{dt}(t; u) + A(t)y(t; u) &= f + Bu, \\ y(0; v) &= y_0. \end{aligned} \right\} \tag{2.1}$$

where $A(t)$ satisfies (1.6)-(1.8),

$$B \in L\big(U : L^2(0, T; V')\big). \tag{2.2}$$

and U is the Hilbert space of control.

Let

$$U_{ad} := \text{ be a closed convex, bounded subset of } U$$

$$K := \text{ be a closed convex subset of } H$$

Let us assume that(approximately controllability)

$$\left.\begin{array}{l} \text{there exists a } u \in U_{ad} \text{ such that for some appropriate} \\[2mm] \tau \in]0, T[, \text{ we have } y(\tau; u) \in K \end{array}\right\} \qquad (2.3)$$

We set

$$\tau^0 = \inf \tau, \quad \tau \text{ satisfying } (2.3) \qquad (2.4)$$

τ^0 termed *optimal time.*

The problem which we shall study are:

(i) existence of an optimal control, that is , existence of $u^0 \in U_{ad}$ such that

$$y(\tau^0; u^0) \in K. \qquad (2.5)$$

(ii) properties of the optimal control ,if it exists.

2.2   Existence theorem

Theorem 2.2.1. *We assume that*(1.6), (1.7) *and* (2.2)- (2.4) *hold. Then there exist an optimal control, that is* $u^0 \in U_{ad}$ *, such that* (2.5) *is satisfied* .

Proof. Let τ_n be such that

$$y(\tau_n; u_n) = y_1, \quad u_n \in U_{ad}, \quad \tau_n \to \tau^0.$$

Set $y_n = y(u_n)$ .Since U_{ad} is bounded , we may verify that y_n (resp. $\frac{dy_n}{dt}$) ranges in a bounded set in $L^2(0, T; V)$(resp. $L^2(0, T; V')$).

We may then extract a subsequence, again denoted by $\{u_n, y_n\}$ such that

$$\left.\begin{array}{l} u_n \to u^0 \quad \text{weakly in } U, \quad u \in U_{ad}, \\[2mm] y_n \to y \quad \text{weakly in } L^2(0, T; V) \\[2mm] \dfrac{dy_n}{dt} \to \dfrac{dy}{dt} \quad \text{weakly in } L^2(0, T; V') \end{array}\right\} \qquad (2.6)$$

We deduce from the equality

$$\frac{dy_n}{dt} + A(t)y_n = f + Bu_n$$

that

$$\frac{dy}{dt} + A(t)y = f + Bu^0$$

$$\text{and} \quad y(0) = y_0.$$

and hence

$$y = y(u^0)$$

But

$$y(\tau_n; u_n) - y(\tau^0; u^0) = y(\tau_n; u_n) - y(\tau^0; u_n) + y(\tau^0; u_n) - y(\tau^0; u^0)$$

Now from (2.6)

$$y(\tau^0; u_n) \to y(\tau^0; u^0) \quad \text{weakly in } H \tag{2.7}$$

and

$$\|y(\tau_n; u_n) - y(\tau^0; u_n)\|_{V'} = \|y(\tau_n; u_n)\|_{V'} = \left\| \int_{\tau^0}^{\tau_n} \frac{d}{dt} y(t; u_n) dt \right\|_{V'}$$

$$\leq \sqrt{\tau_n - \tau^0} \left(\left\| \int_{\tau^0}^{\tau_n} \frac{d}{dt} y(t; u_n) dt \right\|_{V'}^2 dt \right)^{\frac{1}{2}} \tag{2.8}$$

$$\leq c\sqrt{\tau_n - \tau^0}$$

Combine(2.7) and (2.8)show that

$$y(\tau_n; u_n) - y(\tau^0; u^0) \to 0 \quad \text{weakly in } V'.$$

and so, $y(\tau^0; u^0) \in K$ as K is closed and convex, hence weakly closed. This shows that K is reached in time τ^0 by admissible control u^0. $\qquad\square$

2.3   Properties of the optimal control

Theorem 2.3.1. *If the same data and hypotheses of Theorem 2.2.1 hold and the target set K has a special form i.e.*

$$K = \{y \in H : \|y - z\|_H \leq \epsilon\} \tag{2.9}$$

where $z \in H$ and $\epsilon > 0$ are given. Then, the optimal control u^0 is characterized by;

$$\int_0^{\tau^0} \langle p(t), u(t) - u^0(t) \rangle_H \, dt \geq 0 \quad \forall u \in U_{ad}, \tag{2.10}$$

p is the solution of

$$-\frac{dp}{dt}(u^0) + A^*(t)p(u^0) = 0 \quad in]0, \tau^0[,$$
$$p(\tau^0; u^0) = (y(\tau^0; u^0) - z).. \tag{2.11}$$

Proof. First, from Theorem 2.2.1 K is reached in time τ^0 by admissible control u^0. From Remark 1.3.1 Chapter 1, the mapping $t \to y(t; u)$, from $[0, T] \to H$, is continuous for each fixed u and so $y(\tau^0; u) \notin int\, K$, for any $u \in U_{ad}$, by minimality of τ^0.

Using Theorem 1.3.1 it is easy to verify that the mapping $u \to y(\tau^0; u)$, from $U \to H$, is continuous and linear, then, the set

$$\mathcal{A}(\tau^0) = \{y(\tau^0; u) \,:\, u \in U_{ad}\}$$

is the image under a linear mapping of a convex set hence $\mathcal{A}(\tau^0)$ is convex .Thus we have $A(\tau^0) \cap int K = \emptyset$ and $y(\tau^0; u^0) \in \partial K$ (boundary of K) . Since $int K \neq \emptyset (from(2.3))$ so there exists a closed hyperplane separating $A(\tau^0)$ and K containing $y(\tau^0; u^0)$ [?] , i.e there is a nonzero $g \in H' = H$ such as

$$\sup_{y \in A(\tau^0)} \left\langle g, y(\tau^0; u) \right\rangle_H \leq \left\langle g, y(\tau^0; u^0) \right\rangle_H \leq \inf_{y \in K} \left\langle g, y(\tau^0; u) \right\rangle_H \qquad (2.12)$$

From the second inequality in $(2.12), g$ must support the set K at $y(\tau^0; u^0)$ i.e

$$\left\langle g, (y(\tau^0; u) - y(\tau^0; u^0)) \right\rangle_H \geq 0 \quad \forall u \in U \qquad (2.13)$$

Finally,(2.13) may be interpreted as follows : let us introduce the adjoint state p by (2.11),then (2.13) is equivalent to (2.10)

From (2.13) , g must support K at $y(\tau^0; u^0)$; and since H is a Hilbert space , g must be of the form

$$g = \lambda(y(\tau^0; u) - z) \quad \text{for some } \lambda > 0.$$

Dividing the inequality (2.13)by λ gives

$$g = (y(\tau^0; u^0) - z)$$

$\square$

Corollary 2.3.1. *If U_{ad} given by*

$$\left. \begin{aligned} U_{ad} &= \{u \in U : u(t) \in H_{ad}\} \\ H_{ad} &= closed\,, bounded, convex\ subset\ of\ H \end{aligned} \right\} \qquad (2.14)$$

Then the optimal control u^0 exists and satisfies

$$u^0(t) \in \partial H_{ad} = boundary \ of \ H_{ad}, \ for \ almost \ all \ t.(\ bang\text{-}bang \ control) \qquad (2.15)$$

.

Proof. the optimality of u^0 being characterized by $(p(t) = p(t; u^0(t)))$,

$$\int_0^{\tau^0} \langle p(t), u(t) - u^0(t) \rangle_H \, dt \geq 0.$$

From Theorem 2.1 [32], this condition is equivalent to

$$\int_0^{\tau^0} \langle p(t), h - u^0(t) \rangle_H \, dt \geq 0 \ \forall h \in H_{ad}, \ \text{almost all} \, t. \qquad (2.16)$$

But we have

$$p(t) \neq 0 \, \text{almost everywhere} \, . \qquad (2.17)$$

This is true , since if $p(s) = 0$ then the backward uniqueness property implies $p = 0$ in $]s, \tau^0[$. Using Remark 1.1 we can verify that $t \to p(t; u^0)$ is a continues mapping from $[0, \tau^0]$ into H Hence $p(\tau^0; u^0) = 0 = y(\tau^0; u^0) - z$ which leads to a contradiction .Hence(2.17). Finally from (2.16) we option (2.15). $\qquad \square$

Corollary 2.3.2. *Let the hypotheses of corollary2.3.1 hold. IF further H_{ad} is strictly convex, the optimal control u is unique.*

Proof. If u_1 and u_2 are two optimal controls , $\frac{1}{2}(u_1 + u_2)$ is also an optimal control (the set of optimal control is convex)and hence from (2.16) and the strict convexity of H_{ad} we obtain $u_1 = u_2$ almost everywhere. $\qquad \square$

2.4   Applications to Parabolic Systems

In this section, we give some mathematical examples in different type systems . In all of the following mathematical examples we assume that the controllability condition (2.3)and (2.4) are hold.

2.4.1   Time-optimal of parabolic system involving Laplace operator

In this subsection we shall find the set of inequalities defining the time optimal control for the parabolic system involving Laplace operator (1.22).

Let $U = L^2(Q)$ is the space of controls. The state of the system is given by solution of

$$\left.\begin{array}{ll} \dfrac{\partial y(u)}{\partial t} - \Delta y(u) = u, & \text{in } Q, \\[2mm] y(x, 0; u) = y_0(x), & \text{in } \Omega, \\[2mm] y(u) = 0, & \text{on } \Sigma. \end{array}\right\} \qquad (2.18)$$

Let

$$\left.\begin{array}{l} \mathbf{U}_{ad} = \left\{ u \in L^2(Q) : |u(x,t)| \le 1 \right\}, \\[2mm] \mathbf{K} = \left\{ y \in L^2(\Omega) : \|y - z\|_{L^2(\Omega)} \le \epsilon \right\}, \end{array}\right\} \qquad (2.19)$$

where z and ϵ are given such that $z \in L^2(\Omega)$, $\epsilon > 0$.

and let us assume that (2.3)and (2.4)are hold. Then from Theorem2.3.1 and its corollaries , time optimal control u^0 exists and its the solution of the following equations and inequalities :

$$\left.\begin{array}{ll} -\dfrac{\partial p(u^0)}{\partial t} - \Delta p(u^0) = 0, & \text{in } Q, \\[2mm] p(\tau^0; u^0) = y(\tau^0; u^0) - z, & \text{in } \Omega, \\[2mm] p(u^0) = 0, & \text{on } \Sigma, \end{array}\right\}$$

$$\int_0^{\tau^0} \int_\Omega p(u^0)(u - u^0)dxdt \ge 0 \quad \forall u \in U_{ad}.$$

together with (2.18)(with $u = u^0$).Moreover , the optimal control is bang -bang (that is $|u^0(x, t)| = 1$ almost everywhere.).

2.4.2 Time optimal of infinite order Parabolic system

In this subsection, we find the set of inequalities defining the time optimal control for parabolic system involving Operator of infinite order

Let $U = L^2(Q)$ is the space of controls .The state of the system is given by solution of

$$\left.\begin{array}{ll} \dfrac{\partial y(u)}{\partial t} + \left(\displaystyle\sum_{|\alpha|=0}^{\infty} a_\alpha(-1)^\alpha D^{2\alpha} \right) y(u) = u, & \text{,in } Q = \Omega \times]0, T[, \\[4mm] y(x, 0; u) = y_0(x) & \text{,in } \Omega, \\[2mm] y(x, t; u) = 0 & \text{,on } \Sigma = \Gamma \times]0, T[. \end{array}\right\} \qquad (2.20)$$

Let

$$\mathbf{U}_{ad} = \left\{ u \in L^2(Q) : |u(x,t)| \le 1 \right\},$$

$$\mathbf{K} = \left\{ y \in L^2(\Omega) : \|y - z\|_{L^2(\Omega)} \le \epsilon \right\}, \tag{2.21}$$

where z and ϵ are given such that $z \in L^2(\Omega)$, $\epsilon > 0$. and let us assume that are hold. Then from Theorem2.3.1 and its corollaries , time optimal control u^0 exists and its the solution of the following equations and inequalities :

$$\left. \begin{array}{ll} -\dfrac{\partial p(u^0)}{\partial t} + \left(\displaystyle\sum_{|\alpha|=0}^{\infty} a_\alpha(-1)^\alpha D^{2\alpha} \right) p(u^0) = 0, & \text{in } Q, \\[3mm] p(\tau^0; u^0) = y(\tau^0; u^0) - z, & \text{in } \Omega, \\[3mm] p(u^0) = 0, & \text{on } \Sigma. \end{array} \right\}$$

$$\int_0^{\tau^0} \int_\Omega p(u^0)(u - u^0)dxdt \ge 0 \quad \forall u \in U_{ad}$$

together with (2.20)(with $u = u^0$).Moreover , the optimal control is bang -bang .

2.4.3 Time-optimal of coupled systems

In this subsection we shall find the set of inequalities defining the time optimal control for the following coupled system involving Laplace operators .

Let $\mathbf{U} = (L^2(Q))^2$ is the space of controls .For a control $\mathbf{u} = (u_1, u_2) \in (L^2(Q))^2$,the state $\mathbf{y}(\mathbf{u}) = (y_1(\mathbf{u}), y_2(\mathbf{u}))$ of the system is given by the solution of

$$\left. \begin{array}{ll} \dfrac{\partial y_1(\mathbf{u})}{\partial t} - \Delta y_1(\mathbf{u}) + y_1(\mathbf{u}) - y_2(\mathbf{u}) = u_1 & \text{,in } Q, \\[3mm] \dfrac{\partial y_2(\mathbf{u})}{\partial t} + y_1(\mathbf{u}) - \Delta y_2(\mathbf{u}) + y_2(\mathbf{u}) = u_2 & \text{,in } Q, \\[3mm] y_1(x,0:\mathbf{u}) = y_{1,0}(x), \quad y_2(x,0;\mathbf{u}) = y_{2,0}(x), & \text{,in } \Omega, \\[3mm] y_1(\mathbf{u}) = y_2(\mathbf{u}) = 0 & \text{,on } \Sigma. \end{array} \right\} \tag{2.22}$$

Let

$$\mathbf{U}_{ad} = \left\{ \mathbf{u} = (u_i)_{i=1}^2 \in (L^2(Q))^2 : |u_i(x,t)| \le 1 \right\},$$

$$\mathbf{K} = \left\{ \mathbf{y} = (y_i)_{i=1}^2 \in (L^2(\Omega))^2 : \|y_i - z_i\|_{L^2(\Omega)} \le \epsilon \right\}, \tag{2.23}$$

where z_i and ϵ are given such that $z_i \in L^2(\Omega)$, $\epsilon > 0$. and let us assume that (2.3)and (2.4)are hold. Then from Theorem2.3.1 and its corollaries , time optimal control $\mathbf{u}^0 = (u_1^0, u_2^0)$ exists and its the solution of the following equations and inequalities :

$$\left.\begin{array}{ll} -\dfrac{\partial p_1(\mathbf{u}^0)}{\partial t} - \Delta p_1(\mathbf{u}^0) + p_1(\mathbf{u}^0) + p_2(\mathbf{u}^0) = 0 & \text{,in } Q, \\[2mm] -\dfrac{\partial p_2(\mathbf{u}^0)}{\partial t} - p_1(\mathbf{u}^0) - \Delta p_2(\mathbf{u}^0) + p_2(\mathbf{u}^0) = 0 & \text{,in } Q, \\[2mm] p_1(\tau^0; \mathbf{u}^0) = y_1(\tau^0; \mathbf{u}^0) - z_1, \;\; p_2(\tau^0; \mathbf{u}^0) = y_2(\tau^0; \mathbf{u}^0) - z_2, & \text{in } \Omega, \\[2mm] p_1(\mathbf{u}^0) = p_2(\mathbf{u}^0) = 0 & \text{,on } \Sigma. \end{array}\right\}$$

$$\int_0^{\tau^0} \int_\Omega \left[p_1(u^0)(u_1 - u_1^0) + p_2(u^0)(u_2 - u_2^0) \right] dxdt \geq 0 \quad \forall \mathbf{u} = (u_1, u_2) \in U_{ad}$$

together with (2.22)(with $u_1 = u_1^0$, $u_2 = u_2^0$).Moreover , the optimal control is bang -bang .

Chapter 3

Controllability of Parabolic Systems

This chapter is devoted to presented some results concerning with controllability of parabolic with different type of controls(distributed - Neumann - Dirichlet - via initial conditions - pointwise) and different type of observation (distributed - in the final state - boundary - pointwise). .

3.1 Problem statement

We consider the system whose state is given by

$$\left.\begin{array}{c} \dfrac{dy}{dt}(t;u) + A(t)y(t;u) = f + Bu, \\[2mm] y(0;u) = y_0. \end{array}\right\} \tag{3.1}$$

where $A(t)$ satisfies (1.6)-(1.8), $B \in L\big(U : L^2(0,T;V')\big)$. and U is the Hilbert space of control. We also given an observation equation

$$z(u) = Cy(u), \quad C \in L\big(W(0,T;V,V') : \mathcal{H}\big), \quad \mathcal{H} \text{ being a Hilbert space.}$$

Definition 3.1.1. *The system whose state is defined by* (3.1)*is said to be controllable if the observation* $z(u)$ *generates a dense (affine) subspace of the space of observations* $\mathcal{H}$.

3.2   Distributed Control

3.2.1   Observation: $y(u) \in L^2(Q)$

Let $u \in L^2(Q)$, the state is given by the solution of

$$\left. \begin{aligned} \frac{\partial y(u)}{\partial t} - \Delta y(u) &= u, && \text{in } Q, \\ y(x, 0; u) &= y_0(x), && \text{in } \Omega, \\ y(u) &= 0, && \text{on } \Sigma, \end{aligned} \right\} \tag{3.2}$$

and the observation

$$y(u) \in L^2(Q) \tag{3.3}$$

As u ranges over $L^2(Q)$, $y(u)$ generates a dense(affine) subspace of $L^2(Q)$; hence the system is controllable.

let us first remark that by translation we may always reduce the problem of controllability to the case were the system (3.2) with $f = 0$, $y_0 = 0$. We can show quit easily that (3.2) is approximately controllable in $L^2(Q)$ if and only if ., $\{y(u) : u \in L^2(Q)\}$ is dense in $L^2(Q)$. By the Hahn-Banach theorem, this will be the case if

$$\int_Q \psi y(u) dx dt = 0, \quad \psi \in L^2(Q), \tag{3.4}$$

for all $u \in L^2(Q)$, implies that $\psi = 0$.

Let us introduce the adjoint state p by the solution of the following system

$$\left. \begin{aligned} -\frac{\partial p}{\partial t} - \Delta p &= \psi \ \ \text{in } Q, \\ p(x, T) &= 0 && \text{in } \Omega, \\ p(x, t) &= 0 && \text{on } \Sigma. \end{aligned} \right\} \tag{3.5}$$

Multiply the first equation in (3.5) by $y(u)$ and integrate by parts from 0 to T, we obtain the following identity:

$$\begin{aligned} \int_Q \psi y(u) dx dt &= \int_Q \left[-\frac{\partial p}{\partial t} - \Delta p \right] y(u) dx dt \\ &= \int_Q p \left[\frac{\partial y(u)}{\partial t} - \Delta y(u) \right] dx dt = \int_Q p u dx dt = 0 \end{aligned}$$

hence $p = 0$. and hence $\psi = 0$.

3.2.2   Observation: $y(T; u) \in L^2(\Omega)$

Let $u \in L^2(Q)$, the state is defined by (3.2) and the observation is

$$y(x, T; u) \in L^2(\Omega) \tag{3.6}$$

The system (3.2) with observation (3.6) is controllable.

In fact let $\psi \in L^2(\Omega)$ with

$$\int_\Omega \psi y(x, T; u) dx = 0, \quad \forall u \in L^2(Q) \tag{3.7}$$

Introduce the adjoint state p as the solution of the following system

$$\left. \begin{aligned} -\frac{\partial p}{\partial t} - \Delta p &= 0 \quad \text{in } Q, \\ p(x, T) &= \psi(x) &&\text{in } \Omega, \\ p(x, t) &= 0 &&\text{on } \Sigma. \end{aligned} \right\} \tag{3.8}$$

Then

$$0 = \int_Q \left[-\frac{\partial p}{\partial t} - \Delta p \right] y(u) dx dt$$

$$= \int_Q pu \, dx dt - \int_\Omega \psi(x) y(x, T; u)$$

hence $\int_Q pu \, dx dt = 0$, hence $p = 0$. and $\psi = 0$.

3.2.3   Pointwise observation

Let $u \in L^2(Q)$, the state is given by the solution of (3.2) and Let $x^1, ..., x^\mu$ be points of Ω; we assume that the observation is $\{y(x^j, t; u)\}$, $1 \le j \le \mu$ provided we can attach a meaning to this.

Here, $y(u) \in H^{2,1}(Q)$, hence $y(u) \in L^2(0, T; H^2(\Omega))$ and $y(x^j, t)$ has meaning (and " $t \to y(x^j, t)$" $\in L^2(0, T)$) if $H^2(\Omega) \subset C^0(\Omega)$, which is true if (and only if) $\frac{1}{2} - \frac{2}{n} < 0$ i.e $n \le 3$.

Hence we make the standing hypothesis that the dimension $n \le 3$. Then the observation is given by

$$\{y(x^j, t; u)\} \in (L^2(0, T))^\mu. \tag{3.9}$$

As u ranges over $L^2(Q)$, $\{y(x^j, t; u)\}$, $1 \le j \le \mu$ generates a dense(affine) subspace of $(L^2(0, T))^\mu$; hence the system (3.2) with observation (3.9) is controllable.

In fact let $\{\psi^j\} \in (L^2(0,T))^\mu$, $1 \le j \le \mu$ with

$$\sum_{j=1}^{\mu} \int_0^T \psi^j \, y(x^j, t; u)dt = 0, \quad \forall u \in L^2(Q) \tag{3.10}$$

Introduce the adjoint state p as the solution of the following system

$$\left.\begin{array}{ll} -\dfrac{\partial p}{\partial t} - \Delta p = \displaystyle\sum_{j=1}^{\mu} \psi^j \delta(x - x^j) & \text{in } Q, \\[2mm] p(x,T) = 0 & \text{in } \Omega, \\[2mm] p = 0 & \text{on } \Sigma. \end{array}\right\} \tag{3.11}$$

where $\delta(. - x^j)$ is the Dirac delta function at x^j

The problem (3.11) admits a unique solution in $L^2(Q)$ (for example), defined by transposition:

$$\int_Q \left[\frac{\partial \phi}{\partial t} - \Delta \phi\right] p \, dx dt = \sum_{j=1}^{\mu} \int_0^T \psi^j \, y(x^j, t; u)dt,$$

$$\forall \phi \in H^{2,1}(Q), \quad \phi(x,T) = 0, \quad \phi_\Sigma = 0.$$

Let $\phi = y(u)$ Then

$$= \int_Q u p \, dx dt = \sum_{j=1}^{\mu} \int_0^T \psi^j \, y(x^j, t; u)dt$$

hence $\int_Q u p \, dx dt = 0$, hence $p = 0$ and hence $\psi = 0$

3.3 Neumann Boundary control

3.3.1 Observation: $y(u) \in L^2(Q)$

Let $u \in L^2(\Sigma)$, the state is given by the solution of

$$\left.\begin{array}{ll} \dfrac{\partial y(u)}{\partial t} - \Delta y(u) = 0, & \text{in } Q, \\[2mm] y(x,0;u) = y_0(x), & \text{in } \Omega, \\[2mm] \dfrac{\partial y(u)}{\partial \nu} = u, & \text{on } \Sigma, \end{array}\right\} \tag{3.12}$$

and the observation

$$y(u) \in L^2(Q) \tag{3.13}$$

As u ranges over $L^2(\Sigma)$, $y(u)$ generates a dense(affine) subspace of $L^2(Q)$; hence the system (3.12) with observation (3.13) is controllable.

In fact let $\psi \in L^2(Q)$ with

$$\int_Q \psi y(u) dx dt = 0, \quad \forall u \in L^2(\Sigma) \tag{3.14}$$

Introduce the adjoint state p as the solution of the following system

$$\left. \begin{array}{ll} -\dfrac{\partial p}{\partial t} - \Delta p = \psi & \text{in } Q, \\[2mm] p(x,T) = 0 & \text{in } \Omega, \\[2mm] \dfrac{\partial p}{\partial \nu} = 0 & \text{on } \Sigma. \end{array} \right\} \tag{3.15}$$

Then

$$\int_Q \psi y(u) dx dt = \int_Q \left[-\frac{\partial p}{\partial t} - \Delta p \right] y(u) dx dt$$

$$= -\int_\Sigma p \frac{\partial y(u)}{\partial \nu} d\Sigma$$

hence $\int_\Sigma pu d\Sigma = 0$, hence $p = 0$ on Σ. The Cauchy data of p on Σ being zero, we conclude (see [66]) $p = 0$ and hence $\psi = 0$

3.3.2   Observation: $y(T; u) \in L^2(\Omega)$

Let $u \in L^2(\Sigma)$, the state is defined by (3.12) and the observation is

$$y(x, T; u) \in L^2(\Omega) \tag{3.16}$$

The system (3.12) with observation (3.16) is controllable.

In fact let $\psi \in L^2(\Omega)$ with

$$\int_\Omega \psi y(x, T; u) dx = 0, \quad \forall u \in L^2(\Sigma) \tag{3.17}$$

Introduce the adjoint state p as the solution of the following system

$$\left. \begin{array}{ll} -\dfrac{\partial p}{\partial t} - \Delta p = 0 & \text{in } Q, \\[2mm] p(x,T) = \psi(x) & \text{in } \Omega, \\[2mm] \dfrac{\partial p}{\partial \nu} = 0 & \text{on } \Sigma. \end{array} \right\} \tag{3.18}$$

Then

$$0 = \int_Q \left[-\frac{\partial p}{\partial t} - \Delta p \right] y(u)dxdt$$
$$= \int_\Sigma p \frac{\partial y(u)}{\partial \nu} d\Sigma - \int_\Omega \psi(x)y(x,T;u)$$

hence $\int_\Sigma pud\Sigma = 0$, hence $p = 0$ on Σ. The Cauchy data of p on Σ being zero, we conclude (see [66]) $p = 0$ and hence $\psi = 0$

3.3.3   Observation: $y(u) \in L^2(\Sigma)$

Let $u \in L^2(\Sigma)$, the state is given by the solution of (3.12) and the observation

$$y(u) \in L^2(\Sigma) \tag{3.19}$$

As u ranges over $L^2(\Sigma)$, $y(u)$ generates a dense(affine) subspace of $L^2(\Sigma)$; hence the system (3.12) with observation (3.19) is controllable.

In fact let $\psi \in L^2(\Sigma)$ with

$$\int_\Sigma \psi y(u)d\Sigma` = 0, \quad \forall u \in L^2(\Sigma) \tag{3.20}$$

Introduce the adjoint state p as the solution of the following system

$$\left. \begin{array}{ll} -\dfrac{\partial p}{\partial t} - \Delta p = 0 & \text{in } Q, \\[2mm] p(x,T) = 0 & \text{in } \Omega, \\[2mm] \dfrac{\partial p}{\partial \nu} = \psi & \text{on } \Sigma. \end{array} \right\} \tag{3.21}$$

Then

$$0 = \int_Q \left[-\frac{\partial p}{\partial t} - \Delta p \right] y(u)dxdt$$
$$= \int_\Sigma upd\Sigma - \int_\Sigma \psi y(u)d\Sigma$$

hence $\int_\Sigma upd\Sigma = 0$, hence $p = 0$ on Σ. The Cauchy data of p on Σ being zero, we conclude (see [66]) $p = 0$ and hence $\psi = 0$

3.4 Dirichlet Boundary Control

3.4.1 Observation: $y(u) \in L^2(Q)$

Let $u \in L^2(\Sigma)$, the state is given by the solution of

$$\left.\begin{array}{ll} \dfrac{\partial y(u)}{\partial t} - \Delta y(u) = 0, & \text{in } Q, \\[2mm] y(x,0;u) = y_0(x), & \text{in } \Omega, \\[2mm] y(u) = u, & \text{on } \Sigma, \end{array}\right\} \qquad (3.22)$$

The solution of (3.22) is defined by transposition: there exists a unique $y(u) \in L^2(Q)$ such that

$$\int_Q \left[-\frac{\partial \phi}{\partial t} - \Delta \phi \right] y(u)\,dxdt = \int_\Sigma u \frac{\partial \phi}{\partial \nu}\,d\Sigma \quad \forall \phi \in \Phi, \qquad (3.23)$$

where

$$\Phi = \left\{ \phi : \ \phi \in H^{2,1}(Q), \quad \phi(x,T) = 0, \quad \phi_\Sigma = 0 \right\}$$

As u ranges over $L^2(\Sigma)$, $y(u)$ generates a dense(affine) subspace of $L^2(Q)$; hence the system (3.22) with observation $y(u) \in L^2(Q)$ is controllable.

In fact let $\psi \in L^2(Q)$ with

$$\int_Q \psi y(u)\,dxdt = 0, \quad \forall u \in L^2(\Sigma) \qquad (3.24)$$

Introduce the adjoint state p as the solution of the following system

$$\left.\begin{array}{ll} -\dfrac{\partial p}{\partial t} - \Delta p = \psi & \text{in } Q, \\[2mm] p(x,T) = 0 & \text{in } \Omega, \\[2mm] p = 0 & \text{on } \Sigma. \end{array}\right\} \qquad (3.25)$$

In (3.23), we take $\phi = p$. Then

$$\int_Q \psi y(u)\,dxdt = \int_\Sigma u \frac{\partial p}{\partial \nu}\,d\Sigma$$

hence $\int_\Sigma u \frac{\partial p}{\partial \nu} = 0$, hence $\frac{\partial p}{\partial \nu} = 0$ on Σ. The Cauchy data of p on Σ being zero, we conclude (see [66]) $p = 0$ and hence $\psi = 0$

3.4.2 Observation: $y(T;u) \in H^{-1}(\Omega)$

Let $u \in L^2(\Sigma)$, the state is defined by (3.22) and the observation is

$$y(x,T;u) \in H^{-1}(\Omega) \qquad (3.26)$$

Since $y(u) \in L^2(Q) = L^2(0, T; L^2(\Omega))$ (and is defined by (3.23)) and $\Delta y(u) \in L^2(0, T; H^2(\Omega))$, we deduce from (3.22) that $\frac{d}{dt} y(u) \in L^2(0, T; H^2(\Omega))$, from which we may deduce (cf. Lions-Magenes [33] Chapter 1) that $t \to y(t; u)$ is continuous function of $[0, T] \to H^{-1}(\Omega)$. Hence (3.26) has meaning and the observation is in $H^{-1}(\Omega)$

The system (3.22) with observation (3.26) is controllable. To see this, let $\psi \in H_0^1(\Omega)$ such that

$$< y(x, T; u), \psi >= 0, \quad \forall u \in L^2(\Sigma) \tag{3.27}$$

where the bracket denotes duality between $H^{-1}(\Omega)$ and $H_0^1(\Omega)$.

Let us define p by

$$\left. \begin{array}{ll} -\dfrac{\partial p}{\partial t} - \Delta p = 0 & \text{in } Q, \\[2mm] p(x, T) = \psi(x) & \text{in } \Omega, \\[2mm] p = 0 & \text{on } \Sigma. \end{array} \right\} \tag{3.28}$$

Then

$$0 = \int_Q \left[-\frac{\partial p}{\partial t} - \Delta p \right] y(u) dx dt$$
$$= \int_\Sigma y(u) \frac{\partial p}{\partial \nu} d\Sigma - < y(x, T; u), \psi >$$

from which we deduce that $\int_\Sigma u \frac{\partial p}{\partial \nu} d\Sigma = 0$, hence $\frac{\partial p}{\partial \nu} = 0$ on Σ. The Cauchy data of p on Σ being zero, we conclude (see [66]) $p = 0$ and hence $\psi = 0$

3.4.3 Observation: $\frac{\partial y(u)}{\partial \nu} \in H^{-1}(\Sigma)$

Let $u \in L^2(\Sigma)$, the state is given by the solution of (3.22) and the observation $\frac{\partial y(u)}{\partial \nu} \in H^{-1}(\Sigma)$

As u ranges over $L^2(\Sigma)$, $\frac{\partial y(u)}{\partial \nu}$ generates a dense(affine) subspace of $H^{-1}(\Sigma)$; hence the system (3.22) with observation $\frac{\partial y(u)}{\partial \nu} \in H^{-1}(\Sigma)$ is controllable.

In fact let $\psi \in H^{-1}(\Sigma)$ with

$$< \frac{\partial y(u)}{\partial \nu}, \psi >= 0, \quad \forall u \in L^2(\Sigma) \tag{3.29}$$

where the bracket denotes duality between $H^{-1}(\Sigma)$ and $H_0^1(\Sigma)$. Introduce the adjoint state p as the solution of the following system

$$-\frac{\partial p}{\partial t} - \Delta p = 0 \quad \text{in } Q,$$
$$p(x, T) = 0 \qquad\qquad \text{in } \Omega, \left.\right\} \qquad (3.30)$$
$$p = \psi \qquad\qquad \text{on } \Sigma.$$

Then

$$0 = \int_Q \left[-\frac{\partial p}{\partial t} - \Delta p\right] y(u)dxdt$$
$$= \int_\Sigma u\frac{\partial p}{\partial \nu}d\Sigma - < \frac{\partial y(u)}{\partial \nu}, \psi >$$

hence $\int_\Sigma u\frac{\partial p}{\partial \nu}d\Sigma = 0$, hence $\frac{\partial p}{\partial \nu} = 0$ on Σ. The Cauchy data of p on Σ being zero, we conclude (see [66]) $p = 0$ and hence $\psi = 0$

3.5   Control via initial condition

3.5.1   Observation: $y(u) \in L^2(Q)$

Let $u \in L^2(\Omega)$, the state is given by the solution of

$$\frac{\partial y(u)}{\partial t} - \Delta y(u) = 0, \qquad\qquad \text{in } Q,$$
$$y(x, 0; u) = u(x), \qquad\qquad \text{in } \Omega, \left.\right\} \qquad (3.31)$$
$$y(u) = 0, \qquad\qquad \text{on } \Sigma,$$

and the observation $y(u) \in L^2(Q)$

As u ranges over $L^2(\Omega)$, $y(u)$ generates a dense(affine) subspace of $L^2(Q)$; hence the system (3.32) with observation $y(u) \in L^2(Q)$ is controllable. To see this, let $\psi \in L^2(Q$ such that

$$\int_Q \psi y(u)dxdt = 0, \quad \forall u \in L^2(\Omega) \qquad (3.32)$$

Let us define p by

$$-\frac{\partial p}{\partial t} - \Delta p = \psi \quad \text{in } Q,$$
$$p(x, T) = 0 \qquad\qquad \text{in } \Omega, \left.\right\} \qquad (3.33)$$
$$p(x, t) = 0 \qquad\qquad \text{on } \Sigma.$$

Multiply the first equation in (3.33) by $y(u)$ and integrate by parts from 0 to T, we obtain the following identity:

$$\int_Q \psi y(u) dx dt = \int_Q \left[-\frac{\partial p}{\partial t} - \Delta p \right] y(u) dx dt$$

$$= \int_\Omega u p(0) dx$$

hence $p(0) = 0$. But from the backward uniqueness property, $p = 0$ and hence $\psi = 0$

3.5.2 Observation: $y(T; u) \in L^2(\Omega)$

Let $u \in L^2(\Omega)$, the state is given by the solution of (3.32)

As u ranges over $L^2(\Omega)$, $y(T; u)$ generates a dense(affine) subspace of $L^2(\Omega)$; hence the system (3.32) with observation $y(T; u) \in L^2(\Omega)$ is controllable. To see this, let $\psi \in L^2(\Omega$ such that

$$\int_\Omega \psi y(T; u) dx = 0, \quad \forall u \in L^2(\Omega) \tag{3.34}$$

Let us define p by

$$\left. \begin{array}{ll} -\dfrac{\partial p}{\partial t} - \Delta p = 0 & \text{in } Q, \\[2mm] p(x, T) = \psi & \text{in } \Omega, \\[2mm] p(x, t) = 0 & \text{on } \Sigma. \end{array} \right\} \tag{3.35}$$

Multiply the first equation in (3.35) by $y(u)$ and integrate by parts from 0 to T, we obtain the following identity:

$$0 = \int_Q \left[-\frac{\partial p}{\partial t} - \Delta p \right] y(u) dx dt$$

$$= \int_\Omega u p(0) dx - \int_\Omega \psi y(T; u) dx$$

hence $p(0) = 0$. But from the backward uniqueness property, $p = 0$ and hence $\psi = 0$.

3.6   Pointwise Control

Let the state be given by the solution of

$$\left.\begin{array}{ll} \dfrac{\partial y(u)}{\partial t} - \Delta y(u) = \displaystyle\sum_{j=1}^{\mu} u^j(t)\delta(x - x^j), & x^j \in Q, \\[3mm] y(x, 0; u) = y_0(x), & \text{in } \Omega, \\[2mm] y(u) = 0, & \text{on } \Sigma, \end{array}\right\} \quad (3.36)$$

Let $x^1, ..., x^\mu$ be points of Ω; we assume that the observation is $\{y(x^j, t; u)\}$, $1 \leq j \leq \mu$- provided we can attach a meaning to this.

Here, $y(u) \in H^{2,1}(Q)$, hence $y(u) \in L^2(0, T; H^2(\Omega))$ and $y(x^j, t)$ has meaning (and " $t \to y(x^j, t)$" $\in L^2(0, T)$) if $H^2(\Omega) \subset C^0(\Omega)$, which is true if (and only if) $\frac{1}{2} - \frac{2}{n} < 0$ i.e $n \leq 3$.

Hence we make the standing hypothesis that the dimension $n \leq 3$. Then the observation is given by $\{y(x^j, t; u)\} \in (L^2(0, T))^\mu$.

As u ranges over $L^2(Q)$, $\{y(x^j, t; u)\}$, $1 \leq j \leq \mu$ generates a dense(affine) subspace of $(L^2(0, T))^\mu$; hence the system (3.36) with observation $\{y(x^j, t; u)\} \in (L^2(0, T))^\mu$ is controllable.

In fact let $\{\psi^j\} \in (L^2(0, T))^\mu$, $1 \leq j \leq \mu$ with

$$\sum_{j=1}^{\mu} \int_0^T \psi^j\, y(x^j, t; u)dt = 0, \quad \forall u \in L^2(Q) \qquad (3.37)$$

Introduce the adjoint state p as the solution of the following system

$$\left.\begin{array}{ll} -\dfrac{\partial p}{\partial t} - \Delta p = \displaystyle\sum_{j=1}^{\mu} \psi^j \delta(x - x^j) & \text{in } Q, \\[3mm] p(x, T) = 0 & \text{in } \Omega, \\[2mm] p = 0 & \text{on } \Sigma. \end{array}\right\} \qquad (3.38)$$

where $\delta(. - x^j)$ is the Dirac delta function at x^j

The problem (3.38) admits a unique solution in $L^2(Q)$ (for example), defined by transposition:

$$\int_Q \left[\dfrac{\partial \phi}{\partial t} - \Delta \phi\right] pdxdt = \sum_{j=1}^{\mu} \int_0^T \psi^j\, y(x^j, t; u)dt,$$

$$\forall \phi \in H^{2,1}(Q), \quad \phi(x, T) = 0, \quad \phi_\Sigma = 0.$$

Let $\phi = y(u)$ Then

$$= \int_Q u p \, dx \, dt = \sum_{j=1}^{\mu} \int_0^T \psi^j \, y(x^j, t; u) \, dt$$

hence $\int_Q u p \, dx \, dt = 0$, hence $p = 0$ and hence $\psi = 0$

Chapter 4

Time Optimal Control of $n \times n$ Parabolic Lag Systems and Some Regularity Properties

In this chapter, using Lions and Magenes schemes, the time optimal control problem for $n \times n$ parabolic system with constant lags appear both in the parabolic equations and in the Neumann conditions are considered. The existence of a unique regular solution of such $n \times n$ parabolic equations are discussed and the time optimal control is characterized by the adjoint equations. Using this characterization particular properties of the time optimal control are proved.

4.1 Parabolic lag system

In this section , we extend the results given in [2] to the case of $n \times n$ parabolic system in which the control is a vector function.

Let Ω be a bounded open set in $\Re^N$ with smooth boundary Γ. For each time, $t \in]0, T[, T < \infty$, we define a family of bilinear functionals on $(H^1(\Omega))^n$ by

$$\pi(t; ., .) : (H^1(\Omega))^n \times (H^1(\Omega))^n \to \Re,$$

$$\pi(t; \mathbf{y}, \varphi) = \sum_{k=1}^{n} \left[\int_{\Omega} \sum_{i,j=1}^{N} \left(a_{ij}^k(x, t) \frac{\partial y_k}{\partial x_i} \frac{\partial \varphi_k}{\partial x_j} \right) dx + \int_{\Omega} \sum_{j=1}^{n} M_{ij} y_j \varphi_k dx \right]$$

where $\mathbf{y} = (y_k)_{k=1}^n, \varphi = (\varphi_k)_{k=1}^n \in (H^1(\Omega))^n$, the functions $a_{ij}^k(x,t)$ are real $C^\infty-$ functions defined on $\bar{Q}$ (closure of Q)satisfying the ellipticity conditions

$$\sum_{i,j=1}^N a_{ij}^k(x,t)\zeta_i\zeta_j \geq \alpha_k \sum_{i=1}^N \zeta_i^2, \quad \alpha_k > 0, \ \forall(x,t) \in \bar{Q} \ \forall\zeta_i \in \Re, \ 1 \leq k \leq n,$$

and

$$M_{ij} = \begin{cases} I & \text{if } i \geq j \\ -I & \text{if } i < j \end{cases}, \quad I \text{ is the identity operator}$$

$\pi(t; \mathbf{y}, \varphi)$ can be written in the operator form :

$$\begin{aligned}
\pi(t; \mathbf{y}, \varphi) &= \sum_{k=1}^n \int_\Omega \left[\sum_{i,j=1}^N \left(a_{ij}^k(x,t)\frac{\partial y_k}{\partial x_i}\frac{\partial \varphi_k}{\partial x_j} \right) + \sum_{j=1}^n d_{ij}y_j\varphi_k \right] dx \\
&= \sum_{k=1}^n \int_\Omega \left[-\sum_{i,j=1}^N \frac{\partial}{\partial x_j}\left(a_{ij}^k(x,t)\frac{\partial y_k}{\partial x_i} \right)\varphi_k + \sum_{j=1}^n d_{ij}y_j\varphi_k \right] dx \\
&= \sum_{k=1}^n \int_\Omega \left[-\sum_{i,j=1}^N \frac{\partial}{\partial x_j}\left(a_{ij}^k(x,t)\frac{\partial y_k}{\partial x_i} \right) + \sum_{j=1}^n d_{ij}y_j \right]\varphi_k dx \\
&= \sum_{k=1}^n \left\langle A_k(t)y_k + \sum_{j=1}^n d_{ij}y_j, \varphi_k \right\rangle \\
&= \langle \mathbf{A}(t)\mathbf{y}, \varphi \rangle
\end{aligned} \qquad (4.1)$$

i.e

$$\mathbf{A}(t)\mathbf{y} = \begin{pmatrix} A_1(t) & -I & \dots & -I \\ I & A_2(t) & \dots & -I \\ \vdots & \vdots & \dots & \vdots \\ I & I & \dots & A_n(t) \end{pmatrix}_{n \times n} \begin{pmatrix} y_1 \\ y_2 \\ \vdots \\ y_n \end{pmatrix}$$

and $A_k(t)$ are given by

$$A_k(t)y = -\sum_{i,j=1}^N \frac{\partial}{\partial x_i}\left(a_{ij}^k(x,t)\frac{\partial y(x,t)}{\partial x_j} \right) \quad 1 \leq k \leq n$$

It is clear that $\mathbf{A}(t)$ satisfies the coercivity condition

$$\langle \mathbf{A}(t)\mathbf{y}, \mathbf{y} \rangle = \pi(t; \mathbf{y}, \mathbf{y}) \geq \lambda\|\mathbf{y}\|_{(H^1(\Omega))^n}, \quad \lambda = \max\{\alpha_k\}_{k=1}^n$$

Also for all $\mathbf{y}, \varphi \in (H^1(\Omega))^n$ the function $\langle \mathbf{A}(t)\mathbf{y}, \varphi \rangle$ is continuous differentiable with respect to $t \in [0, T]$

Now we consider a system described by the following matrix parabolic equation involving time delay:

$$\left.\begin{array}{l} \dfrac{\partial}{\partial t}\mathbf{y}(\mathbf{u}) + \mathbf{A}(t)\mathbf{y}(\mathbf{u}) = \mathbf{f}(x,t) \quad, (x,t) \in Q =]0,T[\times\Omega \\[2mm] \mathbf{f}(x,t) = \mathbf{b}(x,t)\mathbf{y}(x,t-h) + \mathbf{u}(x,t). \end{array}\right\} \tag{4.2}$$

where $\mathbf{A}(t)$ is given by (4.1), $\mathbf{b} = (b_i)_{i=1}^{n}$ is a given real vector $C^{\infty}-$ function defined on Q; the time delay h is a specified positive number and the vector function $\mathbf{u} = (u_i)_{i=1}^{n}$ corresponds to either a distributed or a specified vector function defined on Q.

We consider the following boundary condition

$$\left.\begin{array}{l} \dfrac{\partial}{\partial \nu_{\mathbf{A}}}\mathbf{y}(x,t) = \mathbf{g}(x,t) \quad \text{on } \Sigma =]0,T[\times\Gamma, \\[2mm] \mathbf{g}(x,t) = \mathbf{c}(x,t)\mathbf{y}(x,t-h) + \mathbf{v}(x,t). \end{array}\right\} \tag{4.3}$$

where

$$\frac{\partial}{\partial \nu_{\mathbf{A}}} = \begin{pmatrix} \frac{\partial}{\partial \nu_{A_1}} & 0 & \cdots & 0 \\ 0 & \frac{\partial}{\partial \nu_{A_2}} & \cdots & 0 \\ \vdots & \vdots & \cdots & \vdots \\ 0 & 0 & \cdots & \frac{\partial}{\partial \nu_{A_n}} \end{pmatrix}_{n \times n}$$

and

$$\frac{\partial y_k}{\partial \nu_{A_k}} = \sum_{i,j=1}^{N} a_{ij}^{k}(x,t)\frac{\partial y_k}{\partial x_j}\cos(\nu, x_j) \quad 1 \le k \le n$$

, $\cos(\nu, x_j) = j - th$ direction cosine of ν, ν being the normal at Γ exterior to Ω, $\mathbf{c} = (c_i)_{i=1}^{n}$ is a given real vector $C^{\infty}-$ function defined on Σ ;$\mathbf{v} = (v_i)_{i=1}^{n}$ represent either a boundary control or a given vector function defined on Σ. The initial data for (4.2) are given by

$$\left.\begin{array}{ll} \mathbf{y}(x,t') = \phi_0(x,t') = (\phi_{i,0}(x,t'))_{i=1}^{n}, & (x,t) \in \Omega\times] - h,0[\\[1mm] \mathbf{y}(x,0) = \mathbf{y}_0(x) = (y_{i,0}(x))_{i=1}^{n}, & x \in \Omega \\[1mm] \mathbf{y}(x,t') = \psi_0(x,t') = (\psi_{i,0}(x,t'))_{i=1}^{n}, & (x,t) \in \Gamma\times] - h,0[\end{array}\right\} \tag{4.4}$$

where $y_{i,0}$, $\phi_{i,0}$, $\psi_{i,0}$ are specified functions. In the sequel , we shall first establish sufficient conditions for the existence of a unique solution of the mixed initial boundary value problem (4.2)-(4.4) then , various optimal control problems will be considered .

4.2   Existence and uniqueness of solutions.

For simplicity, we introduce the following notations:

$$E^j = [(j-1)h, jh[, \quad Q^j = \Omega \times E^j, \quad \Sigma^j = \Gamma \times E^j$$
$$E^0 = [-h, 0[, \qquad Q^0 = \Omega \times E^0, \quad \Sigma^j = \Gamma \times E^0$$

Now ,The existence of a unique solution for the mixed initial -boundary value problem (4.2)-(4.4) , on Q can be established by first solving the problem on Q^1 ,. Then ,the existence of a unique solution on Q^2 . is established by using the solution on Q_1 to generate the initial data at $t = h$. This advancing process is repeated for Q^3, Q^4, ... etc until the procedure covers the whole cylinder Q, Hereafter, the solution on Q^j will be denoted by $\mathbf{y}^j$ $j = 1, 2, \ldots$. Now ,the existence of a unique solution $\mathbf{y}^j = (y_i^j)_{i=1}^n$ can be established by making use the results of Lions and Magenes [33] specialized to the case of a second order parabolic system (4.2)defined on vector functions with boundary condition (4.3) and initial data at $t = (j-1)h$:

$$\mathbf{y}^j(x, (j-1)h) = \mathbf{y}^{j-1}(x, (j-1)h), \quad x \in \Omega. \tag{4.5}$$

Note that once the solution $\mathbf{y}^{j-1}$ defined on Q_{j-1} is determined , the right -hand sides of (4.2)and (4.3)become known functions given by

$$\mathbf{f}^j(x, t) = \mathbf{b}(x, t)\mathbf{y}^{j-1}(x, t-h) + \mathbf{u}(x, t), \tag{4.6}$$

$$\mathbf{g}^j(x, t) = \mathbf{c}(x, t)\mathbf{y}^{j-1}(x, t-h) + \mathbf{v}(x, t), \tag{4.7}$$

For optimal control problems , it is of importance to consider the cases where the controls u_i or v_i belongs to $L^2(Q)$ or $L^2(\Sigma)$ respectively .

Lemma 4.2.1. ($u_i \in L^2(Q)$) (*Lions and Magenes [33]p.33*)

Let $\mathbf{u}$, $\mathbf{f}^j$, $\mathbf{y}^j(., (j-1)h)$ and $\mathbf{g}^j$ be given vectors with

$$\mathbf{u} \in (L^2(Q))^n,$$

$$\mathbf{f}^j \in (L^2(Q^j))^n, \tag{4.8}$$

$$\mathbf{y}^j(., (j-1)h) \in (H^1(\Omega))^n, \tag{4.9}$$

$$\mathbf{g}^j \in H^{\frac{1}{2},\frac{1}{4}}(\Sigma^j)^n. \tag{4.10}$$

Then, there exists a unique solution $\mathbf{y}^j = (y_i^j)_{i=1}^n \in H^{2,1}(Q^j)^n$ *for the mixed initial-boundary value problem* $(4.2),(4.3)$ *and* (4.5) *defined on* Q_j.

Evidently for $j = 1$

$$\mathbf{y}^{j-1}(x, t-h) = \phi_0(x, t-h) \quad in \ Q^0$$

$$\mathbf{y}^{j-1}(x, t-h) = \psi_0(x, t-h) \quad on \ \Sigma^0.$$

and conditions (4.8)and (4.10)can be satisfied if we assume that

$$\phi_0 \in H^{2,1}(Q^0)^n, \quad \psi_0 \in H^{\frac{1}{2},\frac{1}{4}}(\Sigma^0)^n, \text{ and } \mathbf{v} \in H^{\frac{1}{2},\frac{1}{4}}(\Sigma)^n$$

Thus , under the above assumptions, the existence of a unique solution $\mathbf{y}^1 = (y_i^1)_{i=1}^n \in H^{2,1}(Q^1)^n$ follows from Lemma 4.2.1 if $\mathbf{y}_0(x) \in (H^1(\Omega))^n$.

In order to extend the result to Q^2 , it is sufficient to verify that

$$\mathbf{f}^2 \in (L^2(Q^2))^n,$$

$$\mathbf{y}^2(., h) = \mathbf{y}^1(., h) \in (H^1(\Omega))^n,$$

$$\mathbf{g}^2 \in H^{\frac{1}{2},\frac{1}{4}}(\Sigma^2)^n.$$

We note that $\mathbf{y}^1 \in H^{2,1}(Q^1)^n$ implies $\mathbf{y}^1 \in L^2(0, h; (H^2(\Omega))^n) \subset (L^2(Q^1))^n$ and $\frac{\partial \mathbf{y}^1}{\partial t} \in L^2(0, h; (H^0(\Omega)^n))$ then from (4.6) with $j = 2$ and the assumption that $\mathbf{b}$ is C^∞- vector function, if $\mathbf{u} \in (L^2(Q))^n$ then $\mathbf{f}^2 \in (L^2(Q^2))^n$. From the trace theorem [, p.9,Theorem.2.1] [33], $\mathbf{y}^1 \in H^{2,1}(Q^1)^n$ implies that $\mathbf{y}^1 \to \mathbf{y}^1|_{\Sigma^1}$, is a continuous linear mapping of $H^{2,1}(Q^1)^n \to H^{\frac{1}{2},\frac{1}{4}}(\Sigma^1)^n$ then from (4.7) with $j = 2$ and the assumption that $\mathbf{c}$ is C^∞- vector function, if $\mathbf{v} \in H^{\frac{1}{2},\frac{1}{4}}(\Sigma)^n$ then $\mathbf{f}^2 \in H^{\frac{1}{2},\frac{1}{4}}(\Sigma^2)^n$ .From a result of Lions and Magenes [[33] , p.19.Theorem .3.1], $\mathbf{y}^1 \in H^{2,1}(Q^1)^n$,implies that the mapping $t \to \mathbf{y}^1(., h)$ is continuous from $[0, h] \to (H^1(\Omega))^n$ .Consequently , $\mathbf{y}^1(., h) \in (H^1(\Omega))^n$.By Lemma 4.2.1 , there exists a unique solution $\mathbf{y}^2 \in H^{2,1}(Q^2)^n$ .Evidently , we can extend the result to any Q^j, $j = 3, 4, \dots$. The foregoing result is now summarized.

Theorem 4.2.1. *Let* $\mathbf{y}_0$, ϕ_0, ψ_0, $\mathbf{v}$, *and* $\mathbf{u}$ *be given with* $\mathbf{y}_0 \in (H^1(\Omega))^n$, $\phi_0 \in H^{2,1}(Q_0)^n$, $\psi_0 \in H^{\frac{1}{2},\frac{1}{4}}(\Sigma_0)^n$, $\mathbf{v} \in H^{\frac{1}{2},\frac{1}{4}}(\Sigma)^n$ *and* $\mathbf{u} \in (L^2(Q))^n$. *Then , there* exists a unique solution $\mathbf{y} = (y_i)_{i=1}^n \in H^{2,1}(Q)^n$ for problem (4.2)- (4.4).Moreover, $\mathbf{y}(.,jh) \in (H^1(\Omega))^n$ for $j = 1, 2, \ldots$.

Lemma 4.2.2. ($v_i \in L^2(\Sigma)$) (Lions and Magenes [33]p.81)

Let $\mathbf{v}$, $\mathbf{f}^j$, $\mathbf{y}^j(.,(j-1)h)$ and $\mathbf{g}^j$ be given vectors with

$$\mathbf{v} \in (L^2(\Sigma))^n,$$

$$\mathbf{f}^j \in H^{\frac{-1}{2}, \frac{-1}{4}}(Q)^n, \tag{4.11}$$

$$\mathbf{y}^j(.,(j-1)h) \in (H^{\frac{1}{2}}(\Omega))^n, \tag{4.12}$$

$$\mathbf{g}^j \in (L^2(\Sigma_j))^n. \tag{4.13}$$

Then, there exists a unique solution $\mathbf{y}^j = (y_i^j)_{i=1}^n \in H^{\frac{3}{2}, \frac{3}{4}}(Q)^n$ for the mixed initial-boundary value problem (4.2),(4.3)and(4.5)defined on Q_j.

Evidently for $j = 1$,conditions (4.11)-(4.12) can be satisfied if we assume that

$$\phi_0 \in H^{\frac{3}{2}, \frac{3}{4}}(Q_0)^n, \quad \mathbf{y}_0 \in (H^{\frac{1}{2}}(\Omega))^n \text{ and } \psi_0 \in (L^2(\Sigma_0))^n.$$

These assumptions are sufficient to ensure the existence of a unique solution $\mathbf{y}^1 \in H^{\frac{3}{2}, \frac{3}{4}}(Q_1)^n$ To extend the result to Q^2 by means of Lemma 4.2.2, it is sufficient to verify that

$$\mathbf{y}^1 \in H^{\frac{-1}{2}, \frac{-1}{4}}(Q_1)^n$$

$$\mathbf{y}^1(.,h) \in (H^{\frac{1}{2}}(\Omega))^n,$$

$$\mathbf{y}^1|_{\Sigma_1} \in (L^2(\Sigma_1))^n$$

Since $\mathbf{y}^1 \in H^{\frac{3}{2}, \frac{3}{4}}(Q_1)^n$ implies that the mapping $t \to \mathbf{y}^1(.,h)$ is continuous from $[0,T] \to (H^{\frac{3}{4}}(\Omega))^n$ [,p..19,Theorem.3.1 [33]] ,hence $\mathbf{y}^1(.,h) \in (H^{\frac{3}{4}}(\Omega))^n \subset (H^{\frac{1}{2}}(\Omega))^n$. Again from the trace theorem [,p.9,Theorem.2.1 [33]], $\mathbf{y}^1 \in H^{\frac{3}{2}, \frac{3}{4}}(Q_1)^n$ implies that $\mathbf{y}^1 \to \mathbf{y}^1|_{\Sigma_1}$ is a continuous linear mapping of $H^{\frac{3}{2}, \frac{3}{4}}(Q_1)^n \to H^{1, \frac{1}{2}}(\Sigma_1)^n$ Thus, $\mathbf{y}^1|_{\Sigma_1} \in (L^2(\Sigma_1))^n$. We now summarize the foregoing result .

Theorem 4.2.2. Let $\mathbf{y}_0$, ϕ_0, ψ_0, $\mathbf{u}$, and $\mathbf{v}$ be given with $\mathbf{y}_0 \in (H^{\frac{1}{2}}(\Omega))^n$, $\phi_0 \in H^{\frac{3}{2}, \frac{3}{4}}(Q_0)^n$, $\psi_0 \in (L^2(\Sigma_0))^n$, $\mathbf{u} \in H^{\frac{-1}{2}, \frac{-1}{4}}(Q)^n$ and $\mathbf{v} \in (L^2(\Sigma))^n$. Then , there exists a unique solution $\mathbf{y} = (y_i)_{i=1}^n \in H^{\frac{3}{2}, \frac{3}{4}}(Q)^n$ for problem (4.2)- (4.4).Moreover, $\mathbf{y}(.,jh) \in (H^{\frac{1}{2}}(\Omega))^n$ for $j = 1, 2, \ldots$.

We shall consider optimal control problems for system (4.2)-(4.4) in which the control corresponds the either **u** or **v** belonging to

$$\mathbf{u} \in \mathbf{U}_Q = \{\mathbf{u} = (u_i)_{i=1}^n \in (L^2(Q))^n : \quad |u_i(x,t)| \le 1\}$$

or

$$\mathbf{v} \in \mathbf{U}_\Sigma = \{\mathbf{v} = (v_i)_{i=1}^n \in (L^2(\Sigma))^n : \quad |u_i(x,t)| \le 1\}$$

respectively .

4.3   Control problems

Problem 1 :

Let $\mathbf{y}_0$, ϕ_0 , ψ_0 and **v** be given vector functions satisfying the hypothesis of Theorem.4.2.1 Let $\mathbf{y}(t;\mathbf{u})$ denote the solution of (4.2)-(4.4) at (x,t) corresponding to a given control $\mathbf{u} \in \mathbf{U}_Q$ .Occasionally, we write $\mathbf{y}(x,t;\mathbf{u})$ when the explicit dependence on x is required .

In this section ,the minimum-time problem for (4.2)-(4.4) will be considered .Namely , $z_i \in L^2(\Omega)$ and $\epsilon_i > 0$ $i = 1, 2, \ldots n$ are given ; after setting

$$\mathbf{K} = \left\{ \mathbf{y} = (y_i)_{i=1}^n \in (L^2(\Omega))^n : \quad \|y_i - z_i\|_{L^2(\Omega)} \le \epsilon_i \right\}, \tag{4.14}$$

we consider the problem of hitting the target set **K** in minimum time , that is ,minimizing the time t for which $\mathbf{y}(t,\mathbf{u}) \in \mathbf{K}$ and $\mathbf{u} \in \mathbf{U}_Q$.

In order for this problem to be well posed , we assume the following

$$\text{There exists a } \tau > 0 \text{ and } \mathbf{u} \in \mathbf{U}_Q \text{ with } \mathbf{y}(\tau;\mathbf{u}) \in \mathbf{K} \tag{4.15}$$

Then, the following result holds.

Lemma 4.3.1. *If (4.15) holds .* **K** *is reached in minimum time* τ^0 *by an admissible control* $\mathbf{u}^0 \in \mathbf{U}_Q$ *.Moreover,*

$$\int_\Omega (\mathbf{y}(\tau^0;\mathbf{u}^0) - \mathbf{z}) \, (\mathbf{y}(\tau^0;\mathbf{u}) - \mathbf{y}(\tau^0;\mathbf{u}^0)) dx \ge 0 \quad \forall \mathbf{u} \in \mathbf{U}_Q \tag{4.16}$$

where $\mathbf{z} = (z_i)_{i=1}^n$.

To simplify (4.16), we introduce the following adjoint equaton . For every $\mathbf{u} \in \mathbf{U}_Q$, we define $\mathbf{p} = \mathbf{p}(\mathbf{u}) = \mathbf{p}(x, t; \mathbf{u})$ as the solution of

$$-\frac{\partial \mathbf{p}(\mathbf{u})}{\partial t} + \mathbf{A}^*(t)\mathbf{p}(\mathbf{u}) =$$

$$= \begin{cases} -\mathbf{b}(x, t+h)\mathbf{p}(x, t+h; \mathbf{u}), & (x,t) \in \Omega \times]0, \tau^0 - h[\\ 0, & (x,t) \in \Omega \times]\tau^0 - h, \tau^0[\end{cases} \qquad (4.17)$$

with terminal conditions

$$\mathbf{p}(x, \tau^0 : \mathbf{u}) = \mathbf{y}(x, \tau^0; \mathbf{u}) - \mathbf{z}, \quad x \in \Omega, \qquad (4.18)$$

and boundary conditions

$$-\frac{\partial \mathbf{p}(\mathbf{u})}{\partial \nu_{\mathbf{A}^*}}(x, t) = \begin{cases} \mathbf{c}(x, t+h)\mathbf{p}(x, t+h; \mathbf{u}), & (x,t) \in \Gamma \times]0, \tau^0 - h[\\ 0, & (x,t) \in \Gamma \times]\tau^0 - h, \tau^0[\end{cases} \qquad (4.19)$$

where

$$\mathbf{A}^*(t) = \begin{pmatrix} A_1^*(t) & I & \cdots & I \\ -I & A_2^*(t) & \cdots & I \\ \vdots & \vdots & \cdots & \vdots \\ -I & -I & \cdots & A_n^*(t) \end{pmatrix}_{n \times n} , \quad \frac{\partial}{\partial \nu_{\mathbf{A}^*}} = \begin{pmatrix} \frac{\partial}{\partial \nu_{A_1^*}} & 0 & \cdots & 0 \\ 0 & \frac{\partial}{\partial \nu_{A_2^*}} & \cdots & 0 \\ \vdots & \vdots & \cdots & \vdots \\ 0 & 0 & \cdots & \frac{\partial}{\partial \nu_{A_n^*}} \end{pmatrix}_{n \times n}$$

$$A_k^*(t)p_k = -\sum_{i,j=1}^{N} \frac{\partial p_k(\mathbf{u})}{\partial x_j}\left(a_{ji}^k(x, t)\frac{\partial p_k}{\partial x_i}\right),$$

$$-\frac{\partial p_k(\mathbf{u})}{\partial \nu_{A^*}}(x, t) = \sum_{i,j=1}^{N} a_{ji}(x, t)\cos(\nu, x_i)\frac{\partial p_k(\mathbf{u})}{\partial x_j}(x, t),$$

For simplicity, we introduce the following notations:

$$E_{\tau^0}^j = [\tau^0 - jh, \tau^0 - (j-1)h[, \quad Q_{\tau^0}^j = \Omega \times E_{\tau^0}^j, \quad \Sigma_{\tau^0}^j = \Gamma \times E_{\tau^0}^j$$

We observe that for given $\mathbf{z} = (z_i)_{i=1}^n$ and $\mathbf{u} = (u_i)_{i=1}^n$, problem (4.17)-(4.19) can be solved backward in time starting from $t = \tau^0$ by first obtaining the solution $\mathbf{p} = \mathbf{p}^1 = (p_i^1)_{i=1}^n$ on $Q_{\tau^0}^1$ i.e we solve

$$-\frac{\partial \mathbf{p}(\mathbf{u})}{\partial t} + \mathbf{A}^*(t)\mathbf{p}(\mathbf{u}) = 0, \quad (x,t) \in Q_{\tau^0}^1, \qquad (4.20)$$

with terminal condition (4.18) and boundary condition

$$-\frac{\partial \mathbf{p(u)}}{\partial \nu_{\mathbf{A^*}}}(x,t) = 0, \quad (x,t) \in \Sigma_{\tau^0}^1 = \Gamma \times [\tau^0 - h, \tau^0[. \tag{4.21}$$

Having found $\mathbf{p}^1$ we may proceed to solve the problem on $Q_{\tau^0}^2$ backward in time starting with terminal data at $t = \tau^0 - h$: i.e we solve

$$-\frac{\partial \mathbf{p}^2(\mathbf{u})}{\partial t} + \mathbf{A}^*(t)\mathbf{p(u)} = -\mathbf{b}(x,t+h)\mathbf{p}^1(x,t+h;\mathbf{u}), \quad (x,t) \in Q_{\tau^0}^2, \tag{4.22}$$

with

$$\mathbf{p}^2(x,\tau^0-h) = \mathbf{p}^1(x,\tau^0-h), \quad x \in \Omega \tag{4.23}$$

and with boundary condition

$$-\frac{\partial \mathbf{p}^2(\mathbf{u})}{\partial \nu_{\mathbf{A^*}}}(x,t) = \mathbf{c}(x,t+h)\mathbf{p}^1(x,t+h;\mathbf{u}), \quad (x,t) \in \Sigma_{\tau^0}^2,$$

$$\Sigma_{\tau^0}^2 = \Gamma \times]\tau^0 - h, \tau^0 - 2h[. \tag{4.24}$$

Note that the right-hand sides of (4.22)- (4.24) are completely determined once $\mathbf{p}^1$ is Known.This backward process is repeated until the procedure covers the whole cylinder Q

To establish the existence of a unique solution $\mathbf{p}^1$ on $Q_{\tau^0}^1$,we recall from Theorem 4.2.1 that for any $\mathbf{u} = (u_i)_{i=1}^n \in (L^2(Q))^n$, there exists a uniqe solution $\mathbf{y} = (y_i)_{i=1}^n \in H^{2,1}(Q)^n$ and $\mathbf{y}(.,\tau^0;\mathbf{u}) \in (H^1(\Omega))^n$ Thus ,if $\mathbf{z} \in (H^1(\Omega))^n$,then the right-hand side of (4.18) is in $(H^1(\Omega))^n$ .Now ,we may apply Theorem 4.2.1. (with obvious change of variables)to problem (4.17)-(4.19)(with reversed sence of time , i.e ., $t' = \tau^0 - t$.)to establish the existance of a unique solution $\mathbf{p}^1 \in H^{2,1}(Q_{\tau^0}^1)^n$ with $\mathbf{p}^1(.,\tau^0-h) \in (H^1(\Omega))^n$. The result can be extended to $Q_{\tau^0}^2$, and for any $Q_{\tau^0}^j, j = 3, 4, \ldots$, in the same way , since the right-hand side of (4.24) is in $H^{\frac{1}{2},\frac{1}{4}}(\Sigma_{\tau^0}^2)$ (by the trace theorem).Thus ,we have the following result .

Lemma 4.3.2. *Let the hypothesists of Theorem.4.2.1 be satisfied . Then ,for given* $\mathbf{z}$ *in* $(H^1(\Omega))^n$ *and any* $\mathbf{u} \in (L^2(Q))^n$ *, there exists a unique solution* $\mathbf{p(u)} \in H^{2,1}(Q)^n$ *to problem* (4.17)-(4.19).

Now,in view of Lemma.4.3.2 ,we can proceed to simplify (4.16) using the adjoint equation . Setting $\mathbf{u} = \mathbf{u}^0$ in (4.17)-(4.19), multiply (4.17) by $\mathbf{y(u)} - \mathbf{y(u^0)}$ and integrate over $]0,\tau^0[\times\Omega$ we option the identity:

$$\int_\Omega (\mathbf{y}(x, \tau^0; \mathbf{u}^0) - \mathbf{z}), \mathbf{y}(\tau^0, \mathbf{u}) - \mathbf{y}(\tau^0; \mathbf{u}_0) dx$$

$$= \int_0^{\tau^0} \int_\Omega (\mathbf{p}(\mathbf{u}^0) \frac{\partial}{\partial t}(\mathbf{y}(\mathbf{u}) - \mathbf{y}(\mathbf{u}^0)) dx dt$$

$$+ \int_0^{\tau^0} \int_\Omega \mathbf{A}^*(t) \mathbf{p}(\mathbf{u}^0)(\mathbf{y}(\mathbf{u}) - \mathbf{y}(\mathbf{u}^0)) dx dt$$

$$+ \int_0^{\tau^0 - h} \int_\Omega \mathbf{b}(x, t+h) \mathbf{p}(x, t+h; \mathbf{u})(\mathbf{y}(\mathbf{u}) - \mathbf{y}(\mathbf{u}^0)) dx dt$$

(4.25)

The second integral on the right-hand side of (4.25) , in view of Green's formula can be expressed as

$$\int_0^{\tau^0} \int_\Omega \mathbf{A}^*(t) \mathbf{p}(\mathbf{u}^0)(\mathbf{y}(\mathbf{u}) - \mathbf{y}(\mathbf{u}^0)) dx dt =$$

$$\int_0^{\tau^0} \int_\Omega \mathbf{p}(\mathbf{u}^0) \mathbf{A}(t)(\mathbf{y}(\mathbf{u}) - \mathbf{y}(\mathbf{u}^0)) dx dt$$

$$+ \int_0^{\tau^0} \int_\Gamma \mathbf{p}(\mathbf{u}^0) \left(\frac{\partial}{\partial \nu_\mathbf{A}} \mathbf{y}(\mathbf{u}) - \frac{\partial}{\partial \nu_\mathbf{A}} \mathbf{y}(\mathbf{u}^0) \right) d\Gamma dt$$

$$- \int_0^{\tau^0} \int_\Gamma \frac{\partial}{\partial \nu_{\mathbf{A}^*}} \mathbf{p}(\mathbf{u}^0) \left(\mathbf{y}(\mathbf{u}) - \mathbf{y}(\mathbf{u}^0) \right) d\Gamma dt$$

(4.26)

Using (4.4),the second term in the right hand side of (4.26)can be rewrite as

$$\int_0^{\tau^0} \int_\Gamma \mathbf{p}(\mathbf{u}^0) \left(\frac{\partial}{\partial \nu_\mathbf{A}} \mathbf{y}(\mathbf{u}) - \frac{\partial}{\partial \nu_\mathbf{A}} \mathbf{y}(\mathbf{u}^0) \right) d\Gamma dt$$

$$= \int_0^{\tau^0} \int_\Gamma \mathbf{p}(\mathbf{u}^0) \mathbf{c}(\mathbf{x}, \mathbf{t}) \left(\mathbf{y}(x, t-h; \mathbf{u}) - \mathbf{y}(x, t-h; \mathbf{u}^0) \right) d\Gamma dt$$

$$= \int_{-h}^{\tau^0 - h} \int_\Gamma \mathbf{p}(x, t'+h; \mathbf{u}^0) \mathbf{c}(\mathbf{x}, \mathbf{t}) \left(\mathbf{y}(\mathbf{u}) - \mathbf{y}(\mathbf{u}^0) \right) d\Gamma dt'$$

and using (4.2)the above equation can be write as

$$\int_0^{\tau^0} \int_\Gamma \mathbf{p}(\mathbf{u}^0) \left(\frac{\partial}{\partial \nu_\mathbf{A}} \mathbf{y}(\mathbf{u}) - \frac{\partial}{\partial \nu_\mathbf{A}} \mathbf{y}(\mathbf{u}^0) \right) d\Gamma dt$$

$$= \int_0^{\tau^0 - h} \int_\Gamma \mathbf{p}(x, t'+h; \mathbf{u}^0) \mathbf{c}(\mathbf{x}, \mathbf{t'+h}) \left(\mathbf{y}(\mathbf{u}) - \mathbf{y}(\mathbf{u}^0) \right) d\Gamma dt'$$

(4.27)

Using (4.19) the last term in the right hand side of (4.26)can be rewrite as

$$\int_0^{\tau^0} \int_\Gamma \frac{\partial}{\partial \nu_{\mathbf{A}^*}} \mathbf{p}(\mathbf{u}^0) \left(\mathbf{y}(\mathbf{u}) - \mathbf{y}(\mathbf{u}^0) \right) d\Gamma dt$$

$$\int_0^{\tau^0 - h} \int_\Gamma \mathbf{c}(\mathbf{x}, \mathbf{t+h}) \mathbf{p}(x, t+h; \mathbf{u}^0) \left(\mathbf{y}(\mathbf{u}) - \mathbf{y}(\mathbf{u}^0) \right) d\Gamma dt$$

(4.28)

Substituting (4.27)-(4.28)into (4.26) and then (4.26) into (4.25) we have

$$\int_\Omega (\mathbf{y}(x, \tau^0; \mathbf{u}^0) - \mathbf{z}), \mathbf{y}(\tau^0, \mathbf{u}) - \mathbf{y}(\tau^0; \mathbf{u}^0) dx$$

$$= \int_0^{\tau^0} \int_\Omega \mathbf{p}(\mathbf{u}^0) \left[\frac{\partial}{\partial t}(\mathbf{y}(\mathbf{u}) - \mathbf{y}(\mathbf{u}^0)) + \mathbf{A}(t)(\mathbf{y}(\mathbf{u}) - \mathbf{y}(\mathbf{u}^0)) \right] dxdt \qquad (4.29)$$

$$+ \int_0^{\tau^0 - h} \int_\Omega \mathbf{b}(x, t+h)\mathbf{p}(x, t+h; \mathbf{u}^0)(\mathbf{y}(\mathbf{u}) - \mathbf{y}(\mathbf{u}^0)) dxdt$$

Using (4.2),The first term in the right hand side of (4.29)can be rewrite as

$$\int_0^{\tau^0} \int_\Omega \mathbf{p}(\mathbf{u}^0) \left[\frac{\partial}{\partial t}(\mathbf{y}(\mathbf{u}) - \mathbf{y}(\mathbf{u}^0)) + \mathbf{A}(t)(\mathbf{y}(\mathbf{u}) - \mathbf{y}(\mathbf{u}^0)) \right] dxdt =$$

$$= \int_0^{\tau^0} \int_\Omega \mathbf{p}(\mathbf{u}^0)(\mathbf{u} - \mathbf{u}^0) dxdt$$

$$- \int_0^{\tau^0} \int_\Omega \mathbf{p}(\mathbf{u}^0)\mathbf{b}(x, t)(\mathbf{y}(x, t-h; \mathbf{u}) - \mathbf{y}(x, t-h; \mathbf{u}^0)) dxdt$$

$$= \int_0^{\tau^0} \int_\Omega \mathbf{p}(\mathbf{u}^0)(\mathbf{u} - \mathbf{u}^0) dxdt$$

$$- \int_{-h}^{\tau^0 - h} \int_\Omega \mathbf{p}(x, t'+h : \mathbf{u}^0)\mathbf{b}(x, t'+h)(\mathbf{y}(\mathbf{u}) - \mathbf{y}(\mathbf{u}^0)) dxdt'$$

and using (4.2)the above equation can be write as

$$\int_0^{\tau^0} \int_\Omega \mathbf{p}(\mathbf{u}^0) \left[\frac{\partial}{\partial t}(\mathbf{y}(\mathbf{u}) - \mathbf{y}(\mathbf{u}^0)) + \mathbf{A}(t)(\mathbf{y}(\mathbf{u}) - \mathbf{y}(\mathbf{u}^0)) \right] dxdt$$

$$= \int_0^{\tau^0} \int_\Omega \mathbf{p}(\mathbf{u}^0)(\mathbf{u} - \mathbf{u}^0) dxdt \qquad (4.30)$$

$$- \int_0^{\tau^0 - h} \int_\Omega \mathbf{p}(x, t'+h : \mathbf{u}^0)\mathbf{b}(x, t'+h)(\mathbf{y}(\mathbf{u}) - \mathbf{y}(\mathbf{u}^0)) dxdt'$$

Substituting (4.30)into (4.29) ,we obtain

$$\int_\Omega (\mathbf{y}(x, \tau^0; \mathbf{u}^0) - \mathbf{z}), \mathbf{y}(\tau^0, \mathbf{u}) - \mathbf{y}(\tau^0; \mathbf{u}^0) dx =$$

$$= \int_0^{\tau^0} \int_\Omega \mathbf{p}(\mathbf{u}^0)(\mathbf{u} - \mathbf{u}^0) dxdt$$

Hence (4.16) is equivalent to

$$\int_0^{\tau^0} \int_\Omega (\mathbf{p}(\mathbf{u}^0)(\mathbf{u} - \mathbf{u}^0) dxdt \geq 0 \quad \forall \mathbf{u} \in \mathbf{U}_Q$$

We now summarize the foregoing result

Theorem 4.3.1. *We assume that* $(4.15)holds$, $\mathbf{z} \in (H^1(\Omega))^n$ *.Then there exist an optimal control* $\mathbf{u}^0$ *for* (4.2)-(4.4) *which is characterized by*

$$\int_0^{\tau^0} \int_\Omega \mathbf{p}(\mathbf{u}^0)(\mathbf{u} - \mathbf{u}^0)dxdt \geq 0 \quad \forall \mathbf{u} \in \mathbf{U}_Q \tag{4.31}$$

i.e

$$\sum_{i=1}^n \int_0^{\tau^0} \int_\Omega p_i(\mathbf{u}^0)(u_i - u_i^0)dxdt \geq 0 \quad \forall \mathbf{u} = (u_i)_{i=1}^n \in \mathbf{U}_Q \tag{4.32}$$

in particulars case

$$u_i^0(x,t) = -sign(p_i(x,t;\mathbf{u}^0)), \quad (x,t) \in \Omega \times]0, \tau^0[, \quad 1 \leq i \leq n, \tag{4.33}$$

whenever $p_i(\mathbf{u}^0) \neq 0 \forall 1 \leq i \leq n$ *. This property leads to the following result :*

Theorem 4.3.2. *(Bang-Bang theorem) We assume that* $(4.15)holds$ *and the coefficients of the operator* $\mathbf{A}(t)$ *and the functions* b_i, c_i *are analytic in* $\bar{\Omega} \times [0, T]$ *and* Ω *has analytic boundary* Γ *.Then there exists a unique optimal control for* (4.2)-(4.4) *.Moreover, the optimal control is bang -bang ,that is* $|u_i^0(x,t)| \equiv 1$ *almost everywhere, and the unique solution of* (4.2)-(4.4) *,*(4.17)-$(4.19)and$ (4.32).

Proof. The theorem will follow from Theorem4.3.1if we can show that $p_i(x,t) \neq 0$ for almost all $(x,t) \in \Omega \times]0, \tau^0[$. We shall show this fact by contradiction . Therefore , we suppose that

$$\mathbf{p}(x,t) = (p_i)_{i=1}^n(x,t) = 0 \quad \text{for} \quad (x,t) \in E \subset \Omega \times]0, \tau^0[,$$

E non null. Let us denote by k_0 the largest nonnegative integer k such that $\tau^0 - kh > 0$.

Suppose firstly that E has non null intersection with $\Omega \times]\tau^0 - h, \tau^0[$ .If $\tau^0 < h$, the same argument applies to $\Omega \times (0, \tau^0)$ .In the cylinder $\Omega \times]\tau^0 - h, \tau^0[$, $\mathbf{p}(\mathbf{u}^0)$ satisfies

$$-\frac{\partial \mathbf{p}(\mathbf{u}^0)}{\partial t} + \mathbf{A}^*(t)\mathbf{p}(\mathbf{u}^0) = 0 \qquad , (x,t) \in \Omega \times]\tau^0 - h, \tau^0[$$

$$\frac{\partial \mathbf{p}(\mathbf{u}^0)}{\partial \nu_{\mathbf{A}^*}} = 0 \qquad , (x,t) \in \Gamma \times]\tau^0 - h, \tau^0[$$

and so , by ([66]), $\mathbf{p}(\mathbf{u}_0)$ must be analytic in the cylinder $\Omega \times]\tau^0 - h, \tau^0[$. As $\mathbf{p}(\mathbf{u}^0)$ is zero in E , it must be identically zero in $\bar{\Omega} \times]\tau^0 - h, \tau^0[$ .From our earlier remarks

, the mapping $t \to \mathbf{p}(t; \mathbf{u}^0)$ is continuous from $[0, T] \to (L^2(\Omega))^n$, and so

$$p_i(\tau^0; \mathbf{u}^0) = 0 = y_i(\tau^0; \mathbf{u}^0) - z_i$$

contradiction .

We shall now show that the case where $E \cap \Omega \times]0, \tau^0 - kh[$ is non null can be inductively reduced to the above .

Note that , in the cylinder $\bar{\Omega} \times]\tau^0 - 2h, \tau^0 - h[$, $\mathbf{p}(\mathbf{u}^0)$ satisfies

$$-\frac{\partial \mathbf{p}(\mathbf{u}^0)}{\partial t} + \mathbf{A}^*(t)\mathbf{p}(\mathbf{u}^0) + \mathbf{b}(x, t + h)\mathbf{p}(x, t + h; \mathbf{u}^0),$$

$$(x, t) \in \Omega \times]\tau^0 - 2h, \tau^0 - h[\tag{4.34}$$

$$\frac{\partial \mathbf{p}(\mathbf{u}^0)}{\partial \nu_{\mathbf{A}^*}} = \mathbf{c}(x, t + h)\mathbf{p}(x, t + h; \mathbf{u}^0),$$

$$(x, t) \in \Gamma \times]\tau^0 - 2h, \tau^0 - h[. \tag{4.35}$$

We have just shown that , $\mathbf{p}|_\Omega(x, t + h; \mathbf{u}^0)$ and $\mathbf{p}|_\Gamma(x, t + h; \mathbf{u}^0)$ are analytic for $(x, t) \in \Omega \times]\tau^0 - 2h, \tau^0 - h[$ and $(x, t) \in \Gamma \times]\tau^0 - 2h, \tau^0 - h[$ respectively , and so $\mathbf{p}(\mathbf{u}^0)$ must be analytic in $\bar{\Omega} \times]\tau^0 - 2h, \tau^0 - h[$,since (4.34) and (4.35) have analytic coefficients [66] .By induction , $\mathbf{p}(\mathbf{u}^0)$ must be analytic in each cylinder

$$\bar{\Omega} \times]\tau^0 - kh, \tau^0 - (k-1)h[, \quad k = 2, 3, \ldots, k_0, \text{ and } \bar{\Omega} \times]0, \tau^0 - k_0 h[$$

If $\mathbf{p}(\mathbf{u}^0)) = 0$ on $E \cap \Omega \times]\tau^0 - kh, \tau^0 - (k-1)h[$ for some $k = 2, 3, \ldots, k_0$ Then , by analyticity and continuity as before,

$$\mathbf{p}(\mathbf{u}^0)) \equiv 0 \quad \text{for} \quad (x, t) \in \bar{\Omega} \times]\tau^0 - kh, \tau^0 - (k-1)h[\tag{4.36}$$

Substituting (4.36)in to (4.17)and (4.19) gives

$$\frac{\partial \mathbf{p}(\mathbf{u}^0)}{\partial t} + \mathbf{A}^*\mathbf{p}(\mathbf{u}^0) = 0 \quad \text{for} \quad (x, t) \in \Omega \times]\tau^0 - (k-1)h, \tau^0 - (k-2)h[,$$

$$\frac{\partial \mathbf{p}(\mathbf{u}^0)}{\partial \nu_{\mathbf{A}^*}}(x, t) = 0 \qquad \text{for} \quad (x, t) \in \Gamma \times]\tau^0 - (k-1)h, \tau^0 - (k-2)h[.$$

So in the cylinder $\bar{\Omega} \times]\tau^0 - (k-1)h, \tau^0 - (k-2)h[$, $\mathbf{p}(\mathbf{u}^0)$ satisfies

$$\frac{\partial \mathbf{p}(\mathbf{u}^0))}{\partial t} + \mathbf{A}^*\mathbf{p}(\mathbf{u}^0) = 0,$$

$$\frac{\partial \mathbf{p}(\mathbf{u}^0))}{\partial \nu_{\mathbf{A}^*}}(x, t) = 0,$$

$$\mathbf{p}(., \tau^0 - (k-1)h; \mathbf{u}^0)) = 0;$$

consequently, by backward uniqueness [20],

$$\mathbf{p}(\mathbf{u}^0) \equiv 0, \quad \bar{\Omega} \times]\tau^0 - (k-1)h, \tau^0 - (k-2)h[.$$

We can repeat this argument until $\mathbf{p}(\tau^0; \mathbf{u}^0) = 0$,which leads to a contradiction .

With regard to controllability assumption (4.15), we can show quit easily that (4.2)-(4.4) is approximately controllable in $(L^2(\Omega))^n$ in any finite time $T > 0$, i.e ., $\{\mathbf{y}(T; \mathbf{u}) : \mathbf{u} \in (L^2(Q))^n\}$ is dense in $(L^2(\Omega))^n$ .By the Hahn-Banach theorem ,this will be the case if

$$\int_\Omega \bar{\mathbf{z}}(x)\mathbf{y}(x, T; \mathbf{u})dx = 0, \quad \bar{\mathbf{z}} \in (L^2(\Omega))^n, \tag{4.37}$$

for all $\mathbf{u} \in (L^2(Q))^n$, implies that $\bar{\mathbf{z}} = 0$.

Let $\mathbf{p} \in H^{2,1}(Q)^n$ be the unique solution of (4.17)-(4.19) with

$$\mathbf{p}(x, T) = \bar{\mathbf{z}}(x), \quad x \in \Omega.$$

The proof of Theorem 4.3.1 showed that

$$\int_\Omega \bar{\mathbf{z}}(x)(\mathbf{y}(\mathbf{u}) - \mathbf{y}(\bar{\mathbf{u}})dx = \int_0^T \int_\Omega \mathbf{p}(\mathbf{u} - \bar{\mathbf{u}}))dxdt;$$

and so ,if (4.37)holds for all $\mathbf{u} \in (L^2(Q))^n$,then

$$\int_0^T \int_\Omega \mathbf{p}\mathbf{u}dxdt = 0,$$

$\mathbf{u} \in (L^2(Q))^n$ and $\mathbf{p} = 0$ in Q .By continuity,

$$\mathbf{p}(x, T) = \bar{\mathbf{z}}(x) = 0$$

for almost all $x \in \Omega$.   □

Problem 2 :

we can formulate the time optimal control problem for the system (4.2)-(4.4) when the control $\mathbf{v} \in \mathbf{U}_\Sigma$ as follows

Let $\mathbf{y}_0$, ϕ_0 , ψ_0 and $\mathbf{u}$ be given vector functions satisfying the hypothesis of Theorem.4.2.2 Let $\mathbf{y}(x, t; \mathbf{v})$ denote the solution of (4.2)-(4.4) at (x,t) corresponding to a given control $\mathbf{v} = (v_i)_{i=1}^n \in \mathbf{U}_\Sigma$ Let the target set $\mathbf{K}$ given by (4.14)

Let us assume that

$$\left. \begin{array}{l} \text{there exists a } \mathbf{v} \in \mathbf{U}_\Sigma \text{ such that for some appropriate} \\ \tau \in]0, T[, \text{ we have } \mathbf{y}(\tau; \mathbf{v}) \in \mathbf{K} \end{array} \right\} \tag{4.38}$$

We set

$$\tau^0 = \inf \tau, \quad \tau \text{ satisfying } (4.38)$$

τ_0 termed *optimal time*.

The problem is to find $\mathbf{v}^0 \in \mathbf{U}_\Sigma$ such that $\mathbf{y}(x, \tau^0; \mathbf{v}^0) \in \mathbf{K}$

Now ,if $\mathbf{y}_0$, ϕ_0 , ψ_0 and $\mathbf{u}$ satisfying the conditions of Theorem.4.2.2 ,then for each $\mathbf{v} \in \mathbf{U}_\Sigma$, $\mathbf{y}(\mathbf{v}) \in H^{\frac{3}{2}, \frac{3}{4}}(Q)^n$.

Similar to the mean Problem , there exist a $\mathbf{v}^0 \in \mathbf{U}_\Sigma$ such that

$$\mathbf{y}(x, \tau^0; \mathbf{v}^0) \in K ,$$

.Moreover,

$$\int_\Omega (\mathbf{y}(x, \tau^0; \mathbf{v}^0) - \mathbf{z})\,(\mathbf{y}(x, \tau^0; \mathbf{v}) - \mathbf{y}(x, \tau^0; \mathbf{v}^0))dx \geq 0 \quad \forall \mathbf{v} \in \mathbf{U}_\Sigma \qquad (4.39)$$

The above inequality can be simplified by introducing an adjoint equation whose form is identical to (4.17)-(4.19) .From Theorem.4.2.2 , for any $\mathbf{v} \in \mathbf{U}_\Sigma$,there exists a unique solution $\mathbf{y}(\mathbf{v}) \in H^{\frac{3}{2}, \frac{3}{4}}(Q)^n$ with $\mathbf{y}(., \tau^0, \mathbf{v}) \in (H^{\frac{1}{2}}(\Omega))^n$. If $\mathbf{z}_d \in (H^{\frac{1}{2}}(\Omega))^n$,then the right-hand side of (4.18) is in $(H^{\frac{1}{2}}(\Omega))^n$. Similar to mean problem , we can establish the existence of a unique solution $\mathbf{p}(\mathbf{v}) \in H^{\frac{3}{2}, \frac{3}{4}}(Q)^n$ to problem (4.17)-(4.19)

Moreover,(4.39) can be simplified as

$$\int_0^{\tau^0} \int_\Gamma \mathbf{p}(\mathbf{v}^0)(\mathbf{v} - \mathbf{v}^0)d\Gamma dt \geq 0 \quad \forall \mathbf{v} \in \mathbf{U}_\Sigma$$

If the coefficients of the operator $\mathbf{A}(t)$ and the functions b_i , c_i are analytic in $\bar{\Omega} \times [0, T]$ and Ω has analytic boundary Γ .Then ,Using Theorem 4.3.2 we can show that $p_i(x, t) \neq 0$ for almost all $(x, t) \in \Omega \times]0, \tau^0[$ and hence the optimal control is unique and bang -bang ,that is $|u_i^0(x, t)| \equiv 1$ almost everywhere, and the unique solution of (4.2)-(4.4) ,(4.17)-(4.19)and (4.39).

4.4 Special cases

1- If we consider the 2×2 parabolic system with distributed control (Problem 1) and time- lag appears in the equation and in the Neumann condition, then the 2×2

state system becomes

$$\frac{\partial y_1(\mathbf{u})}{\partial t} + A_1 y_1(\mathbf{u}) - y_2(\mathbf{u}) + b_1(x,t) y_1(x, t-h; \mathbf{u}) = u_1 \quad \text{in } Q,$$

$$\frac{\partial y_2(\mathbf{u})}{\partial t} + A_2 y_2(\mathbf{u}) + y_1(\mathbf{u}) + b_2(x,t) y_2(x, t-h; \mathbf{u}) = u_2 \quad \text{in } Q,$$

$$y_1(x, t') = \varphi_{1,0}(x), \quad y_2(x, t') = \varphi_{2,0}(x) \qquad \text{in } Q_0,$$

$$y_1(x, 0) = y_{1,0}(x), \quad y_2(x, 0) = y_{2,0}(x) \qquad \text{in } \Omega,$$

$$\frac{\partial y_1(\mathbf{u})}{\partial \eta_A} = c_1(x,t) y_1(x, t-h; \mathbf{u}) + v_1, \qquad \text{on } \Sigma,$$

$$\frac{\partial y_1(\mathbf{u})}{\partial \eta_A} = c_2(x,t) y_2(x, t-h; \mathbf{u}) + v_2, \qquad \text{on } \Sigma,$$

$$y_1(x, t') = \psi_{1,0}(x), \quad y_2(x, t') = \psi_{2,0}(x) \qquad \text{on } \Sigma_0,$$

the 2×2 adjoint system takes the form

$$-\frac{\partial p_1(\mathbf{u})}{\partial t} + A_1^* p_1(\mathbf{u}) + p_2(\mathbf{u}) + b_1(x, t+h) p_1(x, t+h; \mathbf{u}) = 0$$
$$\text{in } \Omega \times]0, \tau^0 - h[,$$

$$-\frac{\partial p_1(\mathbf{u})}{\partial t} + A_1^* p1(\mathbf{u}) + p_2(\bar{u}) = 0 \quad \text{in } \Omega \times]\tau^0 - h, \tau^0[,$$

$$-\frac{\partial p_2(\mathbf{u})}{\partial t} + A_2^* p_2(\mathbf{u}) - p_1(\mathbf{u}) + b_2(x, t+h) p_2(x, t+h; \mathbf{u}) = 0$$
$$\text{in } \Omega \times]0, \tau^0 - h[,$$

$$-\frac{\partial p_2(\mathbf{u})}{\partial t} + A_2^* p_2(\mathbf{u}) - p_1(\mathbf{u}) = 0 \quad \text{in } \Omega \times]\tau^0 - h, \tau^0[,$$

$$p_1(x, \tau^0; \mathbf{u}) = y_1(x, \tau^0; \mathbf{u}) - z_1(x), \qquad \text{in } \Omega,$$

$$p_2(x, \tau_0; \mathbf{u}) = y_2(x, \tau^0; \mathbf{u}) - z_2(x), \qquad \text{in } \Omega,$$

$$\frac{\partial p_1(\mathbf{u})}{\partial \eta_A} = \begin{cases} c_1(x, t+h) p_1(x, t+h) & \text{on } \Gamma \times]0, \tau^0 - h[, \\ 0 & \text{on } \Gamma \times]\tau^0 - h, \tau^0[, \end{cases}$$

$$\frac{\partial^w p_2(\mathbf{u})}{\partial \eta_A} = \begin{cases} c_2(x, t+h) p_2(x, t+h)) & \text{on } \Gamma \times]0, \tau^0 - h[, \\ 0 & \text{on } \Gamma \times]\tau^0 - h, \tau^0[. \end{cases}$$

Finally, the optimal control $\mathbf{u}^0 = (u_1^0, u_2^0)$ is characterized by

$$\int_0^{\tau^0} \int_\Omega \left[p_1(\mathbf{u}^0)(u_1^0 - u_1) + p_1(\mathbf{u}^0)(u_2^0 - u_2) \right] dx\,dt \geq 0 \quad \forall \mathbf{u} = (u_1, u_2) \in U_Q.$$

2- If we consider the 2×2 parabolic system with distributed control (Problem 1) and time- lag appears in the equation only, then the 2×2 state system becomes

$$\frac{\partial y_1(\mathbf{u})}{\partial t} + A_1 y_1(\mathbf{u}) - y_2(\mathbf{u}) + b_1(x,t) y_1(x, t-h; \mathbf{u}) = u_1 \quad \text{in } Q,$$

$$\frac{\partial y_2(\mathbf{u})}{\partial t} + A_2 y_2(\mathbf{u}) + y_1(\mathbf{u}) + b_2(x,t) y_2(x, t-h; \mathbf{u}) = u_2 \quad \text{in } Q,$$

$$y_1(x,t') = \varphi_{1,0}(x), \quad y_2(x,t') = \varphi_{2,0}(x) \qquad \text{in } Q_0,$$

$$y_1(x,0) = y_{1,0}(x), \quad y_2(x,0) = y_{2,0}(x) \qquad \text{in } \Omega,$$

$$\frac{\partial y_1(\mathbf{u})}{\partial \eta_A} = v_1, \qquad \qquad \text{on } \Sigma,$$

$$\frac{\partial y_1(\mathbf{u})}{\partial \eta_A} = v_2, \qquad \qquad \text{on } \Sigma,$$

the 2×2 adjoint system takes the form

$$-\frac{\partial p_1(\mathbf{u})}{\partial t} + A_1^* p_1(\mathbf{u}) + p_2(\mathbf{u}) + b_1(x, t+h) p_1(x, t+h; \mathbf{u}) = 0$$
$$\text{in } \Omega \times]0, \tau^0 - h[,$$

$$-\frac{\partial p_1(\mathbf{u})}{\partial t} + A_1^* p1(\mathbf{u}) + p_2(\bar{u}) = 0 \qquad \text{in } \Omega \times]\tau^0 - h, \tau^0[,$$

$$-\frac{\partial p_2(\mathbf{u})}{\partial t} + A_2^* p_2(\mathbf{u}) - p_1(\mathbf{u}) + b_2(x, t+h) p_2(x, t+h; \mathbf{u}) = 0$$
$$\text{in } \Omega \times]0, \tau^0 - h[,$$

$$-\frac{\partial p_2(\mathbf{u})}{\partial t} + A_2^* p_2(\mathbf{u}) - p_1(\mathbf{u}) = 0 \qquad \text{in } \Omega \times]\tau^0 - h, \tau^0[,$$

$$p_1(x, \tau^0; \mathbf{u}) = y_1(x, \tau^0; \mathbf{u}) - z_1(x), \qquad \text{in } \Omega,$$

$$p_2(x, \tau_0; \mathbf{u}) = y_2(x, \tau^0; \mathbf{u}) - z_2(x), \qquad \text{in } \Omega,$$

$$\frac{\partial p_1(\mathbf{u})}{\partial \eta_A} = 0 \qquad \qquad \text{on } \Gamma \times]0, \tau^0[,$$

$$\frac{\partial^\omega p_2(\mathbf{u})}{\partial \eta_A} = 0 \qquad \qquad \text{on } \Gamma \times]0, \tau^0[.$$

Finally, the optimal control $\mathbf{u}^0 = (u_1^0, u_2^0)$ is characterized by

$$\int_0^{\tau^0} \int_\Omega \left[p_1(\mathbf{u}^0)(u_1^0 - u_1) + p_1(\mathbf{u}^0)(u_2^0 - u_2) \right] dx dt \geq 0 \quad \forall \mathbf{u} = (u_1, u_2) \in U_Q.$$

3- If we consider the 2×2 parabolic system with distributed control (Problem 1) and time- lag appears in the Neumann condition only, then the 2×2 state system becomes

$$\frac{\partial y_1(\mathbf{u})}{\partial t} + A_1 y_1(\mathbf{u}) - y_2(\mathbf{u}) + b_1(x,t)y_1(x,t-h;\mathbf{u}) = u_1 \quad \text{in } Q,$$

$$\frac{\partial y_2(\mathbf{u})}{\partial t} + A_2 y_2(\mathbf{u}) + y_1(\mathbf{u}) + b_2(x,t)y_2(x,t-h;\mathbf{u}) = u_2 \quad \text{in } Q,$$

$$y_1(x,0) = y_{1,0}(x), \quad y_2(x,0) = y_{2,0}(x) \qquad \text{in } \Omega,$$

$$\frac{\partial y_1(\mathbf{u})}{\partial \eta_A} = c_1(x,t)y_1(x,t-h;\mathbf{u}) + v_1, \qquad \text{on } \Sigma,$$

$$\frac{\partial y_1(\mathbf{u})}{\partial \eta_A} = c_2(x,t)y_2(x,t-h;\mathbf{u}) + v_2, \qquad \text{on } \Sigma,$$

$$y_1(x,t') = \psi_{1,0}(x), \quad y_2(x,t') = \psi_{2,0}(x) \qquad \text{on } \Sigma_0,$$

the 2×2 adjoint system takes the form

$$-\frac{\partial p_1(\mathbf{u})}{\partial t} + A_1^* p1(\mathbf{u}) + p_2(\bar{u}) = 0 \qquad \text{in } \Omega \times]0, \tau^0[,$$

$$-\frac{\partial p_2(\mathbf{u})}{\partial t} + A_2^* p_2(\mathbf{u}) - p_1(\mathbf{u}) = 0 \qquad \text{in } \Omega \times]0, \tau^0[,$$

$$p_1(x, \tau^0; \mathbf{u}) = y_1(x, \tau^0; \mathbf{u}) - z_1(x), \qquad \text{in } \Omega,$$

$$p_2(x, \tau_0; \mathbf{u}) = y_2(x, \tau^0; \mathbf{u}) - z_2(x), \qquad \text{in } \Omega,$$

$$\frac{\partial p_1(\mathbf{u})}{\partial \eta_A} = \begin{cases} c_1(x, t+h)p_1(x, t+h) & \text{on } \Gamma \times]0, \tau^0 - h[, \\ 0 & \text{on } \Gamma \times]\tau^0 - h, \tau^0[, \end{cases}$$

$$\frac{\partial^\omega p_2(\mathbf{u})}{\partial \eta_A} = \begin{cases} c_2(x, t+h)p_2(x, t+h)) & \text{on } \Gamma \times]0, \tau^0 - h[, \\ 0 & \text{on } \Gamma \times]\tau^0 - h, \tau^0[. \end{cases}$$

Finally, the optimal control $\mathbf{u}^0 = (u_1^0, u_2^0)$ is characterized by

$$\int_0^{\tau^0} \int_\Omega \left[p_1(\mathbf{u}^0)(u_1^0 - u_1) + p_1(\mathbf{u}^0)(u_2^0 - u_2) \right] dxdt \geq 0 \quad \forall \mathbf{u} = (u_1, u_2) \in U_Q.$$

Remark 4.4.1. *As in cases 1 and 2 we can construct the problem of 2×2 parabolic system with boundary controls and constant time lag appearing in the Neumann condition only ,and the problem of 2×2 parabolic system with boundary controls and constant time lag appearing in the equation only respectively .*

4-If we neglect the time lag with $n = 2$, then the 2×2 coupled system becomes as

,

$$\frac{\partial y_1(\mathbf{u})}{\partial t} + \Delta y_1(\mathbf{u}) - y_2(\mathbf{u}) = u_1 \qquad \text{in } Q,$$

$$\frac{\partial y_2(\mathbf{u})}{\partial t} + \Delta y_2(\mathbf{u}) + y_1(\mathbf{u}) = u_2 \qquad \text{in } Q,$$

$$y_1(x,0) = y_{1,0}(x), \quad y_2(x,0) = y_{2,0}(x) \quad \text{in } \Omega,$$

$$\frac{\partial y_1(\mathbf{u})}{\partial \eta_A} = v_1, \quad \frac{\partial y_2(\mathbf{u})}{\partial \eta_A} = v_2, \qquad \text{on } \Sigma.$$

The 2×2 adjoint system is

$$-\frac{\partial p_1(\mathbf{u})}{\partial t} + \Delta p_1(\mathbf{u}) + p_2(\mathbf{u}) = 0 \quad \text{in } Q,$$

$$-\frac{\partial p_2(\mathbf{u})}{\partial t} + \Delta p_2(\mathbf{u}) - p_1(\mathbf{u}) = 0 \quad \text{in } Q,$$

$$p_1(x, \tau_0; \mathbf{u}) = y_1(x, \tau^0; \mathbf{u}) - z_1(x), \qquad \text{in } \Omega,$$

$$p_2(x, \tau_0; \mathbf{u}) = y_2(x, \tau^0; \mathbf{u}) - z_2(x), \qquad \text{in } \Omega,$$

$$\frac{\partial p_1(\mathbf{u})}{\partial \eta_A} = 0 \qquad \frac{\partial p_1(\mathbf{u})}{\partial \eta_A} = 0 \qquad \text{on } \Sigma.$$

Finally, the optimal control $\mathbf{u}^0 = (u_1^0, u_2^0)$ is characterized by

$$\int_0^{\tau^0} \int_\Omega \left[p_1(\mathbf{u}^0)(u_1^0 - u_1) + p_1(\mathbf{u}^0)(u_2^0 - u_2) \right] dx dt \geq 0 \quad \forall \mathbf{u} = (u_1, u_2) \in U_S.$$

5-If we take $n = 1$ without coupled and without lag. Then the state system becomes

$$\frac{\partial y(u)}{\partial t} + \Delta y(u) = u, \qquad \text{in } Q,$$

$$y(x,0) = y_0(x) \qquad \text{in} \Omega,$$

$$\frac{\partial y(u)}{\partial \eta_A} = v, \qquad \text{on } \Sigma.$$

The adjoint system is

$$-\frac{\partial p(u)}{\partial t} + \Delta p(u) = 0 \qquad \text{in } Q,$$

$$p(x, \tau^0; u) = y(x, \tau_0; u) - z(x), \qquad \text{in } \Omega,$$

$$\frac{\partial P(u)}{\partial \eta_A} = 0 \qquad \text{on } \Sigma.$$

Finally, the optimal control $\mathbf{u}^0 = u_1^0 = u^0$ is characterized by

$$\int_0^{\tau^0} \int_G p(u^0)(u^0 - u) dx dt \geq 0 \quad \forall u \in U_Q.$$

Chapter 5

Time -Optimal Control Problem for $n \times n$ Co-Operative Parabolic Systems with Control in Initial Conditions

In this chapter, time-optimal control problem for a liner $n \times n$ co-operative parabolic system involving Laplace operator is considered. This problem is, steering an initial state $y(0) = u$, with control u, so that an observation $y(t)$ hitting a given target set in minimum time. First, the existence and uniqueness of solutions of such system under conditions on the coefficients is proved. Afterwards necessary and sufficient conditions of optimality are obtained. Finally a scaler case is given.

5.1 $n \times n$ Co-Operative parabolic Systems

Let $H_0^1(\Omega)$, be the usual Sobolev space of order one which consists of all $\phi \in L^2(\Omega)$ whose distributional derivatives $\frac{\partial \phi}{\partial x_i} \in L^2(\Omega)$ and $\phi_\Gamma = 0$ with the scalar product norm

$$\langle y, \phi \rangle_{H_0^1(\Omega)} = \langle y, \phi \rangle_{L^2(\Omega)} + \langle \nabla y, \nabla \phi \rangle_{L^2(\Omega)}, \quad \text{where } \nabla = \sum_{k=1}^{N} \frac{\partial}{\partial x_k}.$$

We have the following dense embedding chain [70]

$$(H_0^1(\Omega))^n \subseteq (L^2(\Omega))^n \subseteq (H_0^{-1}(\Omega))^n.$$

where $H_0^{-1}(\Omega)$ is the dual of $H_0^1(\Omega)$.

Here and everywhere below the vectors are denoted by bold letters. For $\mathbf{y} = (y_i)_{i=1}^n$, $\phi = (\phi_i)_{i=1}^n \in (H_0^1(\Omega))^n$ and $t \in]0, T[$, let us define a family of continues bilinear forms

$$\pi(t; ., .) : (H_0^1(\Omega))^n \times (H_0^1(\Omega))^n \to \Re \quad \text{by}$$

$$\pi(t; \mathbf{y}, \phi) = \sum_{i=1}^n \int_\Omega [(\nabla y_i)(\nabla \phi_i) - a_i(x, t) y_i \phi_i] \, dx - \sum_{i,j=1}^n \int_\Omega a_{ij}(x, t) y_j \phi_i dx \qquad (5.1)$$

where

$$\left. \begin{array}{l} a_i(x, t) \text{ and } a_{ij}(x, t) \text{ are positive functions in } \quad L^\infty(Q), \\ a_{ij} = 0 \quad \text{when } i = j \quad \text{and } a_{ij} \le \sqrt{a_i a_j} \quad \text{when } i \ne j \end{array} \right\} \qquad (5.2)$$

The bilinear form (5.1) can be but in the operator form:

$$\pi(t; \mathbf{y}, \phi) = \sum_{i=1}^n \int_\Omega [(-\Delta y_i) - a_i(x, t) y_i] \, \phi_i dx - \sum_{i,j=1}^n \int_\Omega a_{ij}(x, t) y_j \phi_i dx$$

$$= \sum_{i=1}^n \langle -(A(t)\mathbf{y})_i, \phi \rangle_{L^2(\Omega)}$$

where $A(t)$ is $n \times n$ matrix operator which maps $(H_0^1(\Omega))^n$ onto $(H_0^{-1}(\Omega))^n$ and takes the form

$$A(t)\mathbf{y} = \begin{pmatrix} \Delta + a_1 & a_{12} & \dots & a_{1n} \\ a_{21} & \Delta + a_2 & \dots & a_{2n} \\ \vdots & \vdots & \vdots & \vdots \\ a_{n1} & a_{n2} & \dots & \Delta + a_n \end{pmatrix} \begin{pmatrix} y_1 \\ y_2 \\ \vdots \\ y_n \end{pmatrix}$$

.

Lemma 5.1.1. *If Ω is a regular bounded domain in R^N, with boundary Γ, and if m is positive on Ω and smooth enough (in particular $m \in L^\infty(\Omega)$,) then the eigenvalue problem:*

$$\left. \begin{array}{ll} -\Delta y = \lambda m(x) y & \text{in } \Omega, \\ y = 0 & \text{on } \Gamma \end{array} \right\}$$

possesses an infinite sequence of positive eigenvalues:

$$0 < \lambda_1(m) < \lambda_2(m) \leq \ldots \lambda_k(m) \ldots; \lambda_k(m) \to \infty, \quad as \ k \to \infty.$$

Moreover $\lambda_1(m)$ *is simple, its associate eigenfunction* e_m *is positive, and* $\lambda_1(m)$ *is characterized by:*

$$\lambda_1(m) \int_\Omega m y^2 dx \leq \int_\Omega |\nabla y|^2 dx \tag{5.3}$$

Proof. See[65].   □

Now, let

$$\lambda_1(a_i) \geq n, \quad i = 1, 2, \ldots \ldots, n \tag{5.4}$$

Lemma 5.1.2. *If* (5.2) *and* (5.4) *hold then, the bilinear form* (5.1) *satisfy the Gårding inequality*

$$\pi(t; \mathbf{y}, \mathbf{y}) + c_0 \|y\|^2_{(L^2(\Omega))^n} \geq c_1 \|y\|^2_{(H^1_0(\Omega))^n}, \quad c_0, c_1 > 0$$

Proof. In fact

$$\pi(t; \mathbf{y}, \mathbf{y}) = \sum_{i=1}^n \int_\Omega \left[|\nabla y_i|^2 - a_i(x, t) y_i^2 \right] dx - \sum_{i,j=1}^n \int_\Omega a_{ij}(x, t) y_i y_j dx$$

$$\geq \sum_{i=1}^n \int_\Omega \left[|\nabla y_i|^2 - a_i(x, t) y_i^2 \right] dx - 2 \sum_{i>j}^n \int_\Omega \sqrt{a_i(x, t) a_j(x, t)} y_i y_j dx$$

By Cauchy Schwarz inequality and (5.3),we obtain

$$\pi(t; \mathbf{y}, \mathbf{y}) \geq \sum_{i=1}^n \left(1 - \frac{1}{\lambda_1(a_i)} \right) \int_\Omega |\nabla y_i|^2 dx$$

$$- 2 \sum_{i>j}^n \int_\Omega \left(\frac{1}{\sqrt{\lambda_1(a_i)\lambda_1(a_j)}} \right) \left(\int_\Omega |\nabla y_i|^2 dx \right)^{\frac{1}{2}} \left(\int_\Omega |\nabla y_j|^2 dx \right)^{\frac{1}{2}}$$

$$\geq \sum_{i=1}^n \left(\frac{\lambda_1(a_i) - n}{\lambda_1(a_i)} \right) \int_\Omega |\nabla y_i|^2 dx$$

From (5.4) we have

$$\pi(t; \mathbf{y}, \mathbf{y}) \geq \alpha \left[\sum_{i=1}^n \int_\Omega |\nabla y_i|^2 dx \right] \quad \alpha > 0$$

Add $\|\mathbf{y}\|_{(L^2(\Omega))^n}$ to two sides, then we have the result.   □

We can now apply Theorem 1.1 and Theorem 1.2 Chapter3 in [32] (with $V = (H_0^1(\Omega))^n$ and $H = (L^2(\Omega))^n$) to obtain the following theorem:

Theorem 5.1.1. *If* (5.2) *and* (5.4) *hold, then there exist a unique solution*

$$\mathbf{y} \in W(0,T) = \left\{ \mathbf{y} : \mathbf{y} \in L^2(0,T;(H_0^1(\Omega))^n), \quad \frac{\partial \mathbf{y}}{\partial t} \in (H_0^{-1}(\Omega))^n) \right\}$$

satisfying the following $n \times n$ *system:* $i = 1,2,...,n$

$$\left. \begin{array}{ll} \dfrac{\partial y_i}{\partial t} = (A(t)\mathbf{y})_i + f_i, \quad f_i \in L^2(0,T;H_0^{-1}(\Omega)) & in\ Q, \\[2mm] y_i(x,0) = u_i(x), \quad u_i(x) \in L^2(\Omega) & in\ \Omega, \\[2mm] y_i(x,t) = 0 & on\ \Sigma, \end{array} \right\} \qquad (5.5)$$

Moreover $\mathbf{y}$ is continuous from $[0,T] \rightarrow (L^2(\Omega))^n$

5.2   Minimum time and controllability

We denote the unique solution of (5.5), at time t for each control $\mathbf{u} = (u_1, u_2, ..., u_n)$ by $\mathbf{y}(t; \mathbf{u})$ Occasionally, we write $\mathbf{y}(x,t;\mathbf{u})$ when the explicit dependence on x is required. We can now formulate the time optimal control problem corresponding to the $n \times n$ co-operative parabolic system (5.5):

$$\min\{t: \ \mathbf{y}(t;\mathbf{u}) \in K_\varepsilon^n, \ \mathbf{u} \in U_\epsilon^n \}, \qquad (5.6)$$

with constraints

$$\left. \begin{array}{l} \mathbf{y}(t;\mathbf{u}) \text{ is the solution of } (5.5), \\[2mm] U_\epsilon^n = \{\mathbf{u} = (u_1, u_2, ..., u_n) \in (L^2(\Omega))^n : \ \|u_i\|_{L^2(\Omega)} \leq \epsilon \}, \\[2mm] K_\varepsilon^n = \{\mathbf{z} = (z_1, z_2, ..., z_n) \in (L^2(\Omega))^n : \|z_i - z_{id}\|_{L^2(\Omega)} \leq \varepsilon \}, \end{array} \right\} \qquad (5.7)$$

and $\epsilon,\ \varepsilon > 0$ and $z_{id} \in L^2(\Omega)$ are given.

Theorem 5.2.1. *If* (5.2) *and* (5.4) *are hold, then the system whose state is given by* (5.5) *is controllable,*

i.e.,   there exists a $\tau \in]0,T]$ *and* $\mathbf{u} \in U_\epsilon^n$ *with* $\mathbf{y}(\tau;\mathbf{u}) \in K_\varepsilon^n$ \qquad (5.8)

Proof. let us first remark that by translation we may always reduce the problem of controllability to the case were the system (5.5) with $f_i = 0$. We can show quit easily that (5.5) is approximately controllable in $(L^2(\Omega))^n$ in any finite time $\tau > 0$, if and only if ., $\{\mathbf{y}(\tau; \mathbf{u}) : \mathbf{u} \in (L^2(\Omega))^n\}$ is dense in $(L^2(\Omega)^n$. By the Hahn-Banach theorem, this will be the case if

$$\int_\Omega \bar{z}_i(x) y_i(x, \tau; \mathbf{u}) dx = 0, \quad \bar{z}_i \in L^2(\Omega), \tag{5.9}$$

for all $\mathbf{u} \in (L^2(\Omega))^n$, implies that $\bar{z}_i(x) = 0$, $i = 1, 2, ..., n$.

Let us introduce the adjoint state $p(t; \mathbf{u})$ by the solution of the following system

$$\left.\begin{aligned} -\frac{\partial p_i}{\partial t}(t; \mathbf{u}) - (A^*(t)\mathbf{p}(t; \mathbf{u}))_i &= 0 \quad &\text{in } \Omega \times]0, \tau[, \\ p_i(x, \tau) &= \bar{z}_i(x) \quad &\text{in } \Omega, \\ p_i(x, t) &= 0 \quad &\text{in } \Gamma \times]0, \tau[, \end{aligned}\right\} \tag{5.10}$$

where $A^*(t)$ is the adjoint of $A(t)$ which is defined by

$$\langle A^*(t)\phi, \psi \rangle = \langle \phi, A(t)\psi \rangle, \quad \phi, \psi \in (H_0^1(\Omega))^n.$$

The existence of a unique solution for the Problem (5.10) can be proved using Theorem5.1.1 with an obvious change of variables).

Multiply the first equation in (5.10) by $y_i(t; \mathbf{u})$ and integrate by parts from 0 to τ, we obtain the following identity:

$$\begin{aligned} 0 &= \int_0^\tau \int_\Omega \left[-\frac{\partial p_i}{\partial t} - (A^*(t)\mathbf{p}(t; \mathbf{u}))_i\right] y_i(t; \mathbf{u}) dx dt \\ &= -\int_\Omega p_i(t; \mathbf{u}) y_i(t; \mathbf{u}) \Big|_0^\tau dx + \int_0^\tau \int_\Omega p_i(t; \mathbf{u}) \left[\frac{\partial}{\partial t} y_i(t; \mathbf{u}) - (A(t)\mathbf{y}(t; \mathbf{u}))_i\right] dx dt \\ &= \int_\Omega \bar{z}_i(x) y_i(x, \tau; \mathbf{u}) dx - \int_\Omega p_i(0; \mathbf{u}) u_i dx. \end{aligned}$$

and so, if (5.9)holds, then

$$\int_\Omega p_i(x, 0; \mathbf{u}) u_i dx = 0 \quad \forall u_i \in L^2(\Omega)$$

hence $p_i(x, 0; \mathbf{u}) = 0$. But from the backward uniqueness property, $\mathbf{p} = (p_i)_{i=1}^n \equiv 0$ and hence $\bar{z}_i(x) = 0$.

$\square$

Now set

$$\tau^0 = \inf\{\tau : \mathbf{y}(\tau; \mathbf{u}) \in K_\varepsilon^n \text{ fore some } \mathbf{u} \in U_\epsilon^n\}. \tag{5.11}$$

Then , the following result holds .

Theorem 5.2.2. *If* (5.2) *and* (5.4) *are hold, then there exist an admissible control* $\mathbf{u}^0$ *to the problem* (5.6)-(5.11), *which steering* $\mathbf{y}(t; \mathbf{u}^0)$ *to hitting a target set* K_ε^n *in minimum time* τ^0 *(defined by* (5.11) *). Moreover*

$$\sum_{i=1}^n \int_\Omega \left(y_i(\tau^0; \mathbf{u}^0) - z_{id}\right)\left(y_i(\tau^0; \mathbf{u}) - y_i(\tau^0; \mathbf{u}^0)\right) dx \geq 0 \quad \forall \mathbf{u} \in U_\epsilon^n \tag{5.12}$$

Now Inequality (5.12) can be interpreted as follows: let us introduce the adjoint state $p(t; \mathbf{u}^0)$ by the solution of the following system

$$\left. \begin{array}{ll} -\dfrac{\partial p_i}{\partial t}(t; \mathbf{u}^0) + \left(A^*(t)\mathbf{p}(t; \mathbf{u}^0)\right)_i = 0 & \text{in } \Omega \times]0, \tau^0[, \\[2mm] p_i(x, \tau^0) = (y_i(x, \tau^0) - z_{id}) & \text{in } \Omega, \\[2mm] p_i(x, t) = 0 & \text{in } \Gamma \times]0, \tau^0[, \end{array} \right\} \tag{5.13}$$

As the proof of Theorem 5.2.1, we multiply the first equation int (5.13) by $y_i(t; \mathbf{u}) - y_i(t; \mathbf{u}^0)$ and integrate by parts from 0 to τ^0, we obtain the following identity:

$$\int_\Omega (y_i(\tau^0; \mathbf{u}^0) - z_{id})(y_i(x, \tau^0; \mathbf{u}) - y_i(x, \tau^0; \mathbf{u}^0))dx = \int_\Omega p_i(0; \mathbf{u}^0)(\mathbf{u} - \mathbf{u}^0)dx.$$

hence condition (5.12) becomes

$$\sum_{i=1}^n \int_\Omega p_i(x, 0; \mathbf{u}^0)(\mathbf{u} - \mathbf{u}^0)dx \geq 0 \quad \forall \mathbf{u} \in U_\epsilon^n. \tag{5.14}$$

Using controllability condition (5.8), the backward uniqueness property implies $p_i(x, 0; \mathbf{u}^0) = 0$. then the optimal control is bang -bang,i.e, $||u_i^0||_{L^2(\Omega)} = \epsilon$ and since U_ϵ^n is strictly convex, then the optimal control is unique. We have thus proved:

Theorem 5.2.3. *If* (5.2) *and* (5.4) *are hold, then there exist the adjoint state*

$$\mathbf{p} \in W(0, \tau^0) = \left\{\mathbf{p} : \mathbf{p} \in L^2(0, \tau^0; (H_0^1(\Omega))^n), \quad \dfrac{\partial \mathbf{p}}{\partial t} \in (H_0^{-1}(\Omega))^n)\right\}$$

such that the optimal control $\mathbf{u}^0$ *of problem* (5.6)- (5.11)*is bang -bang unique and it is determined by* (5.13),(5.14) *together with* (5.5) *(with* $u_i = u_i^0$, $i = 1, 2, ..., n$ *).*

5.3   Scaler case

Here, we take the case where $n = 2$, in this case, the time optimal problem therefore is

$$\min\{t : \; y(x,t;\mathbf{u}) \in K_\varepsilon^2, \; \mathbf{u} = (u_1, u_2) \in U_\epsilon^2 \}$$

The state $\mathbf{y} = (y_1, y_2))$ is solution of the following equations

$$\left.\begin{array}{ll}
\dfrac{\partial y_1}{\partial t} - \Delta y_1 = a_{11}(x,t)y_1 + a_{12}(x,t)y_2 + f_1, & x \in \Omega, \quad t \in]0, \tau^0[, \\[2mm]
\dfrac{\partial y_2}{\partial t} - \Delta y_2 = a_{21}(x,t)y_1 + a_{22}(x,t)y_2 + f_2, & x \in \Omega, \quad t \in]0, \tau^0[, \\[2mm]
y_1(x,0) = u_1^0(x), \quad y_2(x,0) = u_2^0(x), & x \in \Omega, \\[2mm]
y_1(x,t) = y_2(x,t) = 0, & x \in \Gamma, \quad t \in]0, \tau^0[,
\end{array}\right\}$$

with

$$\left.\begin{array}{l}
a_{ij}(x,t), \; i,j = 1,2 \; \text{ are positive functions in } \; L^\infty(Q), \\[2mm]
\lambda_1(a_{11}) \geq 2, \quad \lambda_1(a_{22}) \geq 2.
\end{array}\right\}$$

The adjoint is solution of the following equations

$$\left.\begin{array}{ll}
-\dfrac{\partial p_1}{\partial t} - \Delta p_1 = a_{22}(x,t)p_1 - a_{12}(x,t)p_2 + f_1, & x \in \Omega, \quad t \in]0, \tau^0[, \\[2mm]
-\dfrac{\partial p_2}{\partial t} - \Delta p_2 = -a_{21}(x,t)p_1 + a_{11}(x,t)p_2 + f_2, & x \in \Omega, \quad t \in]0, \tau^0[, \\[2mm]
p_1(x, \tau^0) = (y_1(x, \tau^0) - z_{1d}), & x \in \Omega, \\[2mm]
p_2(x, \tau^0) = (y_2(x, \tau^0) - z_{2d}), & x \in \Omega, \\[2mm]
p_1(x,t) = p_2(x,t) = 0, & x \in \Gamma, \quad t \in]0, \tau^0[.
\end{array}\right\}$$

The maximum condition is

$$\int_0^{\tau^0} \int_\Omega \left[p_1(x,0;u^0)(u_1 - u_1^0) + p_1(x,0;u^0)(u_1 - u_1^0) \right] dx\, dt \geq 0 \quad \forall u \in U_\epsilon^2.$$

.

5.4   Comments

- We note that, in this paper, we have chosen to treat a special systems involving Laplace operator, just for simplicity. Most of the results we described in this paper apply, without any change on the results, to more general parabolic systems involving the following second order operator :

$$L(x,.) = \sum_{i,j=1}^{n} b_{ij}(x,.)\frac{\partial^2}{\partial x_i \partial x_j} + \sum_{j=1}^{n} b_j(x,.)\frac{\partial}{\partial x_j} + b_0(x,.)$$

with sufficiently smooth coefficients (in particular, b_{ij}, b_j, $b_0 \in L^{\infty}(Q)$, b_j, $b_0 > 0$) and under the Legendre-Hadamard ellipticity condition

$$\sum_{i,j=1}^{n} \eta_i \eta_j \geq \sigma \sum_{i=1}^{n} \eta_i \quad \forall (x,t) \in Q,$$

for all $\eta_i \in \Re$ and some constant $\sigma > 0$.

In this case, we replace the first eigenvalue of the Laplace operator by the first eigenvalue of the operator L (see [65]).

- In this paper, we have chosen to treat a co-operative parabolic systems with Dirichlet boundary conditions.The results can be extended to the case of $n \times n$ co-operative parabolic system with Neumann boundary conditions: If we take $H^1(\Omega)$ instead of $H_0^1(\Omega)$, we have to replace the Dirichlet boundary conditions $y_i = 0$, $p_i = 0$ on the boundary by Neumann boundary conditions $\frac{\partial y_i}{\partial \nu} = 0$, $\frac{\partial p_i}{\partial \nu} = 0$ where ν is the outward normal.

- The results in this paper, carry over to the fixed -time problem ([32] chapter 3)

$$\text{minimize } \sum_{i=1}^{n} \int_{\Omega} |y_i(x,T;\mathbf{u}) - z_{id}(x)|^2 dx, \quad T \text{ fixed },$$

subject to (5.5) [except in the trivial case where $z_{id}(x) = y_i(x,T;\mathbf{u}) \forall i = 1,2,..,n$ for some admissible control $\mathbf{u}$] This can proven in an analogous manner, as the necessary and sufficient conditions for optimality for this problem coincide with (5.5),(5.10) and (5.14) (with $u_i = u_i^0$, $i = 1,2,...,n$).

Bibliography

[1] Wang , G. A bang-bang principle of time optimal internal controls of heat equation. *arXiv: math/0612237 V1[math.OC]* (2006).

[2] Knowles ,G.,, " Time optimal control of parabolic systems with boundary condition involving time delays " Journal of Optimiz.Theor. Applics . ,**25**, 563-574, (1978).

[3] Henry Hermes and Joseph Lasalle,P. Functional Analysis and Time Optimal Control*Mathematics to Science and Engineering 56, Academic Press New York and Landon 1969,* (1992).

[4] Knowles,G.,An Introduction to Applied Optimal Control *Academic Press, New York,.* (1981)

[5] Fattorini, H.O. The time optimal control problem in Banach spaces. *Applied Mathematics and Optimization 1,* 163-188, (1974).

[6] Fattorini, H.O. The Time Optimal Problem for Distributed Control of Systems Described by the Wave Equation. In: Aziz, A.K., Wingate, J.W., Balas, M.J. (eds.): Control Theory of Systems Governed by Partial Differential Equations.*Academic Press, New York, San Francisco, London* (1957).

[7] Fattorini,H.O. Time-optimal control of solutions of operational differential equations.*SIAM Journal on Control,3,* 54-59, (1964).

[8] Fattorini, H.O. Ordinary differential equations in linear topological spaces. *II Jour. Diff. Equations, 6,* 50-70 (1969).

[9] Fattorini, H.O. Optimal control problems for distributed parameter systems in Banach spaces *Applied Mathematics and Optimization,28*, 225-257, (1993).

[10] Fattorini, H.O. The maximum principle for linear infinite dimensional control systems with state constriants. *Discrete and Continuous Dynamical Systems, 1*, 77-101 (1995).

[11] Fattorini, H.O. Infinite Dimensional Optimization Theory and Opotimal Control. *Cambridge University. Prees*, (1998).

[12] Fattorini, H.O. Some remarks on the time Optimal control problem in infinite dimension. *CHAMPAN and HALL/CRC 411, Calculus of Variations and Optimal Control*, 77-96, (1999).

[13] Fattorini, H.O." The maximum principle for control systems described by linear parabolic equations. *Journal of Mathematical Analysis and Applications, 259*, 630-651, (2001).

[14] Fattorini, H.O." Optimal control of diffusions. *Applied Mathemtics OptimizationJ, 46*, 207-230, (2002).

[15] Wang , P.K.C. Optimal control of parabolic systems with boundary conditions involving time delays. *SIAMJ.Control 13*, 274 - 293, (1975).

[16] Wang , P.K.C. Time-optimal control of time-lag systems with time-lag control. *Journal of Mathematical Analysis and Applications Vol.52,No,3* 366- 378, (1975).

[17] Barbu , V.,Optimal Control of Variational Inequalities *Pitman Advanced Publishing Program Boston . London . Melbourne* (1983).

[18] Yong , J. Time optimal control for distributed parameter systems-existence theory and necessary conditions. *Koda Math Journal 14*, 239-253, (1991).

[19] Li Xunjing and Yao Yunlong Time optimal control of distributed parameter systems. *Journal Scientia Sinica Vol XXIV No 4* , 455-465,(1981).

[20] Friedman , A.,Optimal control for parabolic variational inequalities. *SIAM Journal of Control and Optimization* , 25,482-497, (1987).

[21] Balakrishnan ,A.V., Applied Functional Analysis *Spring Verlag*, (1975).

[22] Balakrishnan,A.V. Optimal control problems in Banach spaces. *SIAM Journal on Control,3*, 152-180, (1965).

[23] Ahmed,N.U., Optimal control of generating policies in a power system goverened by a second order hyperbolic partial differential equation *SIAM Journal Control and Optimization, 15*,1016- 1033, (1977).

[24] Sadek, I.S. Optimal Control of Time-Delay Systems with Distributed Parameters *Journal of optimization theory and applications 67,3* 567585, (1990).

[25] Sloss, J.M., Sadek, I.S.,Bruch, Jr,J.C. and Adali,S. Optimal Control of Structural Dynamic Systems in One Space Dimension Using a Maximum Principle *Journal of Vibration and Control, 11:* 245261, (2005).

[26] Arada,N. & Raymond,J.P., Time optimal problems with Dirichlet boundary condtrols *Discrete and Continuous Dynamical systems 9,4*, 1-20, (2003).

[27] El-Saify , H.A. Optimal control of nxn parabolic system involving time lag *IMA Journal of Mathematical Control and Information 22*,240-250, (2005).

[28] El-Saify , H.A. Quadratic optimal control problem for differential system of Petrowsky type *Dynamics of Continuous, Discrete and Impulsive Systems, Series A: mathematical Analysis 13* 153-163, (2006).

[29] El-Saify , H.A. Optimality conditions for infinite order hyperbolic differential system *Journal of Dynamical and Control systems 12,3* 313-330, (2006).

[30] El-Saify , H.A. Optimal control problem for infinite order parabolic lag system *Journal of Interdisciplinary Mathematics 9*, 1 ,99-114 (2006).

[31] J.-L. Lions, Exact controllability, stabilizability and perturbations for distributed systems, SIAM Review, 30 (1988), 1-68.

[32] J.L.Lions, Optimal control of systems governed by partial differential equations, Springer-verlag, Band **170**, (1971).

[33] J.L.Lions and E.Magenes, " Non homogeneous boundary value problem and applications, **I, II**, Spring-Verlage, New York, (1972).

[34] Lions. J.L. and Zuazua, E. Approximate controllability of a hydroelastic coupled system *ESAM: Control , Optimization and Calculus of Variations 1*, 1-15, (1995).

[35] Shehata, M.A Some Time Optimal Control Problems *M.Sc. Thesis, Faculty of Science, Egept, Beni-Suef University, Mathematics Department.* (2007).

[36] Kotarski, W., El-Saify, H.A.,Shehata,M.A. Time optimal control of parabolic lag system with infinite variables. *Journal of the Egyptian mathematical society 15*, 21-34, (2007).

[37] Kotarski, W., El-Saify, H.A. and Shehata,M.A.,"Minimum time problem for infinite order parabolic lag system" *International Conference on Mathematical Analysis and Its Applications (ICMAA06)* Assuit - Egypt , 3-6 January (2006).

[38] El-Saify , H.A., Shehata,M.A. Time-optimal control of Petrowsky systems with infinitely many variables and control-state constraints *Studies in Mathematical Sciences* Vol. 2, No. 1, pp. 21-35, (2011).

[39] El-Saify , H.A., Shehata,M.A. Time-Optimal Control problem for parabolic systems involving different types of operators. *Journal of the Egyptian mathematical society 15(3),* 405-423,(2009)

[40] H.A. El-Saify, H.M. Serag and M.A.Shehata, Time-optimal control for cooperative hyperbolic systems Involving Laplace operator. *Journal of Dynamical and Control systems.* **15**,3,(2009),405-423.

[41] M.A.Shehata, Some time-optimal control problems for $n \times n$ co-operative hyperbolic systems with distributed or boundary controls. *Journal of Mathematical Sciences: Advances and Applications.* vol 18, No 1-2,(2012),63-83.

[42] M.A.Shehata, Time -optimal control problem for $n \times n$ co-operative parabolic systems with control in initial conditions, Advances in Pure Mathematics Journal , 3, No 9A,(2013),38-43.

[43] M.A.Shehata, Dirichlet Time-Optimal Control of Co-operative Hyperbolic Systems *Advanced Modeling and Optimization Journal.* Volume 16, Number 2, (2014),355-369.

[44] Byung Soo Lee, Mohammed Shehata, Salahuddin , Time -optimal control problem for $n \times n$ co-operative parabolic systems with strong constraint control in initial conditions, Journal of Science and Technology, Vol.4 No.11,(2014)..

[45] D.L. Russell, Controllability and stabilizability theory for linear partial differential equations. Recent progress and open questions, SIAM Review 20 (1978), 639-739.

[46] H. T. BANKS, M. Q. JACOBS and R. M. LATINA, The synthesis of optimal controls for linear, time optimal problems with retarded controls, J . Optimization Theory Appl. 8 (1971), 319360.

[47] J.L.Lions , " Remarks on approximate controllability, J.Anal.Math, **59**, (1992),103-116..

[48] M. C. DELFOUR and S. K. MITTER, Controllability and observability for infinite- dimensional systems, SIAM J. Control Optim. 10 (1972), 329333.

[49] H. O. FATTORINI, Some remarks on complete montrollability, SIAM J. Control Optim. 4 (1966), 686694.

[50] H. O. FATTORINI, Boundary control of temperature distributions in a parallelepipedan, SIAM J. Control Optim. 13 (1975), 113.

[51] H. O. FATTORINI and D. L. RUSSELL, Exact controllability theorems for linear parabolitic equations in one space dimension, Arch. Rational Mech. Anal. 43 (1971), 272292.

[52] J. KLAMKA, Controllability of linear systems with time-variable delays in control, Informat. J . Control 24 (1976), 869878.

[53] J. KLAMKA, Absolute controllability of linear systems with time-variable delays in control, Systems Sci. 4 (1978), 4352.

[54] J. JLLAMKA, Relative controllability of infinite-dimensional systems w ith delays in control, Systems Sci. 4 (1978)) 4352.

[55] A. W. OLBROT, On the controllability of linear systems with time delays in control, IEEE Trans. Automat. Control 17 (1972), 664666.

[56] D. L. RUSSELL, A unified boundary controllability theory for hyperbolic and parabolic partial differential equations, Studies in Appl. Math. 52 (1973), 189 211.

[57] Y. SAKAWA, Controllability for partial differential equations of parabolic type, SIAM J. Control Optim. 12 (1974), 389400.

[58] T. I. SEIDMAN, Observation and prediction for the heat equation. IV: Patch observability and controllability, SIAM J. Control Optim. 15 (1977), 412427.

[59] G. Knowles Time optimal control of parabolic systems with boundary condition involving time delays, Journal of Optimiz.Theor. Applics, **25**,(1978), 563-574 .

[60] R. TRIGGIANI, Controllability and observability in Banach space with bounded operators, SIAM J. Control Optim. 13 (1975), 462491.

[61] R. TRIGGIANI, Extensions of rank conditions for controllability and observability to Banach space and unbounded operators, SIAM J. Control Optim. 14 (1976), 313338.

[62] R. TRIGGIANI, On the lack of exact controllability for mild solutions in Banach space, J. Math. Anal. Appl. 50 (1975), 438446.

[63] H. O. Fattorini, Infinite dimensional Optimization Theory and Opotimal Control, Cambridge Univ. Prees (1998).

[64] Triebel,H. Interpolation Theory. Function Spaces. Differential Operators,*North-Holland, Amsterdam.* (1978).

[65] J. Fleckinger, J. Hernández and F.DE. Thélin, On the existence of multiple principal eigenvalues for some indefinite linear eigenvalue problems, Rev.R.Acad.Cien.Serie A.Mat. 97,2 (2003), 461-466.

[66] H. Tanabe, On differentiability and analyticity of eighted elliptic boundary value problems, *Osaka Math.Journal ,2* (1965),163-190.

[67] K. D. Phung, G. Wang, X. Zhang, On the existence of time optimal controls for linear evolution equations, Discrete and Continuous Dynamical Systems, Ser. B, 8(2007), 925-941.

[68] H. O. Fattorini, Infinite Dimensional Linear Control Systems: The Time Optimal and Norm Optimal Problems, North-HollandMathematics Studies, 201, Elsevier, Amsterdam, (2005).

[69] X. Li, J. Yong, Optimal Control Theory for Infinite Dimensional Systems, Birkhauser Boston, (1995).

[70] Adams, R.A., Sobelev Spaces *Academic Prees , New York,* (1975)

[71] Ahmed Zabel & Maryam Alghamdi., Gardings inequality for elliptic differential operator with infinite Number of Variables, *International Journal of Mathematics and Mathematical Sciences, Vol 2011,* 1-10, (2011).

[72] Artola ,M.," Equations parabolique a retardement *Compts Rendus de l' Academic des Scines, Paris, France,264,*668-671, (1967).

[73] Bellman,R., Glicksberg,I. and Gross,O. On the "bang-bang" control problem. *Quart. Appl. Math. 14,* 11-18, (1956).

[74] Berezanskii, Ju.M., Self-Adjoint Operators in Spaces of Functions of Infinitely Many Variables", *Translation.of Mathematical .Monographs 63,* (1986).

[75] Dubinskii ,Ju.A. " Sobolev spaces of infinite order and behavior of solution of some boundary value problems with unbounded increase of the order of equation." Math.USSR .Sb. **27** ,(1975),143-162 .

[76] Dubinskii ,Ju.A. " Non-triviality of Sobolev spaces of infinite order for a full Euclidean space and a Tours ." Math.USSR .Sb. **29** ,(1976),393- 401 .

[77] Dubinskii ,Ju.A. " About one method for solving partial differential equation ." Doklady Akademii Nauk SSSR, **258**, ,(1981),780-784 .

[78] Dubinskii ,Ju.A. "**Sobolev Spaces of Infinite Order and Differential Equations**" .Mathematics and its applications East European Series **3** (1986).

[79] Fleckinger,J. Hernandez,J. and de Thelin,F. On the existence of multiple principal eigenvalues for some indefinite linear eigenvalue problems._Rev. R. Acad. Cien. Ser. A. Mat. 97,2,_ 461466, (2003).

[80] Lagnese , J., " Boundary Stabilization of Thin Plates. " _Stud . Appl . Math , 10,_ (1989).

Printed by Books on Demand GmbH, Norderstedt / Germany